THE ILLUSTRATED HISTORY OF THE
OTTAWA TENT COLONY

JIM RIDINGS

America Through Time is an imprint of Fonthill Media LLC
www.through-time.com
office@through-time.com

Published by Arcadia Publishing by arrangement with Fonthill Media LLC
For all general information, please contact Arcadia Publishing:
Telephone: 843-853-2070
Fax: 843-853-0044
E-mail: sales@arcadiapublishing.com
For customer service and orders:
Toll-Free 1-888-313-2665

www.arcadiapublishing.com

First published 2021

Copyright © Jim Ridings 2021

ISBN 978-1-63499-334-0

Typeset in Sabon LT Std
Printed and bound in England

Contents

ACKNOWLEDGMENTS

Information came from: LaSalle County Historical Museum in Utica, Amanda Carter and Jennifer Nordstrom; Reddick Library in Ottawa, Laura Youngstrum; United Auto Worker's Pat Greathouse Education Center in Ottawa, Brad Dutcher, UAW Region 4 assistant director; Chicago Public Library, Josh Mabe; Edward-Elmhurst Health, Keith Hartenberger; Newspaper microfilm archives from the *Ottawa Free Trader*, *Ottawa Republican-News*, *The Daily Times* of Ottawa, *The Times* of Ottawa, *Chicago Tribune*, *Chicago Daily News*, *Ottawa Journal*, *Streator Times*, *Bloomington Pantagraph*, *Rockford Republic*, *New York Times*, and *Washington Post*.

Also Phil Walker, Rich Myers, Sharon Schrepple Page, Linda Willibey, Kelly Magoonaugh, Judy Corcoran Dubach, Nancee Duesdieker, Dale Maley, Darlene Halm, Mollie Perrot, Kathalyn Anne Holz-Zammuto, Diane Whalen, and Katie Troccoli. I want to thank my wife, Janet, for her continued support of my work.

Other research sources include: American Lung Association; U.S. National Library of Medicine; *Report by the Commission on Chicago Landmarks*, 2018; National Institutes of Health; Division of Global Public Health, School of Medicine, University of California at San Diego; *American Journal of Respiratory and Critical Care Medicine*; *Wisconsin Medical Journal*; American Clinical and Climatological Association; *Ottawa Old and New*, 1914; *Ottawa Sesquicentennial 1987*; Illinois State Medical Society; *Smithsonian Magazine*, May 2016; *Respiratory Medicine*, August 2006; *The Forward*; Centers For Disease Control (CDC); World Health Organization (WHO); Wikipedia; *Scientific American*, 2007; *American Review of Tuberculosis*; Hodgkins Fund, *Atmospheric Air in Relation to Tuberculosis*, by Dr. Guy Hinsdale, Smithsonian Collection, 1914; American Lung Association; *A History of the National Tuberculosis Association* by Dr. S. Adolphus Knopf; Dr. William Osler, *The Medical News*, December 12, 1903; Dr. William Osler, *On the Study of Tuberculosis*, *The Philadelphia Medical Journal*, December 1900; Adelaide Dutcher, *Where the Danger Lies in Tuberculosis*, *The Philadelphia Medical Journal*, December 1900; Jo Baeza, *The Arizona Independent*, 2011; *Medicine in Territorial Arizona*, Frances E. Quebbeman, Arizona Historical Foundation, 1966; *Journal of Preventive Medicine and Hygiene*, 2017; *American Review of Tuberculosis*; Mayo Clinic online; University of Nebraska; PBS, *The American Experience*; *American Journal of Respiratory and Critical Care Medicine* 2004; *American Association for Respiratory Care*; *Annals of Internal Medicine*; and Colorado Springs Pioneers Museum.

1

THE WHITE DEATH

It was called the White Death. Phthisis. Consumption. Tuberculosis.

It was called the "white death" because it turned its victims so pale. Hippocrates named it "phthisis," meaning significant weight loss or wasting away. It was historically called "consumption" because it ravaged and consumed the body. The sufferers were called "consumptives" and "tuberculars" and sometimes were known as "lungers" in casual references and newspapers.

Tuberculosis is fatal in most people unless treated; however, with drugs, it can be cured if diagnosed in time. Unlike influenza or COVID-19, which are caused by a virus, tuberculosis is a bacterial infection. Tuberculosis attacks the lungs, although the bacteria can attack any part of the body, such as the kidney, spine, and brain.

It cannot be overstated how widespread and deadly tuberculosis has been throughout human history. For centuries, it was the leading cause of death in the world. Tuberculosis is estimated to have killed more than a billion people in the past two centuries, according to the Centers for Disease Control (C.D.C.), the World Health Organization (W.H.O.), and other medical authorities.

That is right—1 billion deaths in just the past 200 years.

You can add hundreds of millions more deaths because tuberculosis has been known to mankind since ancient times. The tuberculosis bacterium has been found in mummified humans as far back as 3,400 B.C.

Tuberculosis pandemics from the 1600s to the 1800s wiped out half the population of Europe several times. In everyday life over the centuries, it killed untold numbers of people.

The C.D.C., W.H.O., and other medical sites have said that by the dawn of the nineteenth century, tuberculosis had killed one in seven of all the people that had ever lived.

Tuberculosis is still a worldwide pandemic. Dr. Anthony Fauci, director of the National Institute of Allergy and Infectious Diseases, said in 2018 that tuberculosis was the leading infectious cause of death worldwide, killing approximately 1.6 million people worldwide in 2017 and 1.4 million people in 2019. In 2018, there were 1.7 billion people infected by the T.B. bacteria—approximately 23 percent of the world's population. In 2019, thirty countries accounted for 87 percent of new T.B. cases.

Consumption

¶ There is no specific for consumption. Fresh air, exercise, nourishing food and Scott's Emulsion will come pretty near curing it, if there is anything to build on. Millions of people throughout the world are living and in good health on one lung.

¶ From time immemorial the doctors prescribed cod liver oil for consumption. Of course the patient could not take it in its old form, hence it did very little good. They *can* take

SCOTT'S EMULSION

and tolerate it for a long time. There is no oil, not excepting butter, so easily digested and absorbed by the system as cod liver oil in the form of Scott's Emulsion, and that is the reason it is so helpful in consumption where its use must be continuous.

¶ We will send you a sample free.

¶ Be sure that this picture in the form of a label is on the wrapper of every bottle of Emulsion you buy.

Scott & Bowne
Chemists
409 Pearl Street
New York

soe. and £1; all druggists

THE ILLUSTRATED LONDON NEWS

DR. DE JONGH'S

(Knight of the Order of Leopold of Belgium and of the Legion of Honour)

LIGHT-BROWN COD-LIVER OIL.

Incontestably proved by thirty years' medical experience to be

THE PUREST, THE MOST PALATABLE, THE MOST DIGESTIBLE, AND THE MOST EFFICACIOUS IN CONSUMPTION, THROAT AFFECTIONS AND DEBILITY OF ADULTS AND CHILDREN.

SELECT MEDICAL OPINIONS.

SIR HENRY MARSH, BART., M.D.,
Physician in Ordinary to the Queen in Ireland.
"I consider Dr. De Jongh's Light-Brown Cod-Liver Oil to be a very pure Oil, not likely to create disgust, and a therapeutic agent of great value."

DR. GRANVILLE, F.R.S.,
Author of "The Spas of Germany."
"Dr. De Jongh's Light-Brown Cod-Liver Oil does not cause the nausea and indigestion too often consequent on the administration of the Pale Oils."

DR. EDGAR SHEPPARD,
Professor of Psychological Medicine, King's College.
"Dr. De Jongh's Light-Brown Cod-Liver Oil has the rare excellence of being well borne and assimilated by stomachs which reject the ordinary Oils."

SIR G. DUNCAN GIBB, BART. M.D.,
Physician to the Westminster Hospital.
"The value of Dr. De Jongh's Light-Brown Cod-Liver Oil as a therapeutic agent in a number of diseases, chiefly of an exhaustive character, has been admitted by the world of medicine."

DR. SINCLAIR COGHILL,
Physician to the Hospital for Consumption, Ventnor.
"In Tubercular and the various forms of Strumous Disease, Dr. De Jongh's Oil possesses greater therapeutic efficacy than any other Cod-Liver Oil with which I am acquainted."

DR. HUNTER SEMPLE,
Physician to the Hospital for Diseases of the Throat.
"I have found Dr. De Jongh's Light-Brown Cod-Liver Oil very useful in cases of Chronic Cough, and especially in Laryngeal Disease complicated with Consumption."

Sold ONLY in Capsuled Imperial Half-Pints, 2s. 6d.; Pints, 4s. 9d.; Quarts, 9s.; by all Chemists and Druggists. Sole Consignees—ANSAR, HARFORD, and CO., 210, High Holborn, London.
CAUTION.—*Resist mercenary attempts to recommend or substitute inferior kinds.*

There have always been quack cures for every ailment. Here are a few advertisements from the 1800s.

Eight countries account for two-thirds of the total—India, Indonesia, China, the Philippines, Pakistan, Nigeria, Bangladesh, and South Africa, according to W.H.O.

In the past two centuries, T.B. has claimed more lives than malaria, influenza, smallpox, H.I.V./A.I.D.S., cholera, and the plague combined. It is mostly under control in the United States (taking 542 lives in America in 2019) due to a number of actions, such as the discovery of antibiotics, better education, and better treatment.

In recent decades, tuberculosis has been somewhat forgotten and over-shadowed by other deadly epidemics, such as the coronavirus COVID-19; H.I.V. (human immunodeficiency virus), which leads to A.I.D.S. (acquired immunodeficiency syndrome); ebola; H1N1 (a coronavirus); S.A.R.S. (severe acute respiratory syndrome, also a coronavirus); and several deadly strains of influenza. However, tuberculosis has been responsible for more deaths than all of those diseases combined.

A consensus was formed in the medical community in the last half of the nineteenth century that the open-air treatment was the best method, at least for patients in the early stage of the disease. The idea was that fresh air was best for the diseased lungs. Fresh air, along with a heavy amount of nutritious food to build up the body, rest, and sunlight, was the method of treatment being used. Drugs had not been developed to cure tuberculosis, and so various methods of treatment were all the doctors could do. Sanitariums were established, many using the "open-air" method of treating patients.

The modern-day crusade started when the Pennsylvania Society for the Prevention of Tuberculosis was formed in 1892, and it gained full steam with the formation of the National Association for the Study and Prevention of Tuberculosis in 1904.

Ottawa, Illinois, was one of the main centers in America where a concerted effort was made to fight this disease.

James Wiley Pettit had a successful medical practice in Ottawa in the 1870s and 1880s. He contracted a case of tuberculosis and left town for several years to recover. When he returned and continued his practice, he took a special interest in tuberculosis.

At the time, it was believed that a hot, dry climate was most important for people suffering from this lung disease. Sufferers went to New Mexico, Arizona, Texas, Colorado, and California for a cure. However, Dr. Pettit heard different opinions while attending the meeting of the International Medical Congress in Rome, Italy, in 1900. He returned to Europe in 1902 to investigate tuberculosis sanatoria and their locations. He saw that whether patients were being treated in the Swiss Alps with cold air and a high altitude or on the Mediterranean with its warm weather at sea level, the results were similar. Dr. Pettit observed the same results in America—whether it was in the cold Adirondacks, in Colorado's high altitude, in New Mexico's dry air, or California's warm weather at sea level. He concluded that climate was not a factor and patients could recover in Illinois as well as anywhere. It was important that patients were treated in their home climate, he said.

One method of treatment was the tent colony where patients were kept outdoors all year round to breathe fresh air. There is a scientific basis for this therapy. Fresh air does prevent T.B. from spreading as rapidly as it otherwise would through the lungs. Even indoors, an open window provides the natural ventilation that moves the air, especially through hospital wards.

This was not as radical as it seems. It was being used in sanitariums across America and around the world. Benjamin Franklin in 1786 extolled the benefits of sleeping with a blanket to keep warm while keeping the windows open to let in fresh air, even on the coldest nights. The flow of fresh air and avoidance of stale air in a closed room was important for health, he said. The idea is that fresh air and ventilation dilutes the bacteria in the air.

Medical science still gives credence to this therapy. An article in *Scientific American* in 2007, about tuberculosis patients in Peru, began, "It turns out that helping prevent the spread of tuberculosis may be as simple as opening a window." It continued, "Researchers found that natural ventilations moves more than twice as much air as through hospital wards than expensive fans." It said the W.H.O. recommended natural ventilation to limit T.B. transmission in poor areas, and it cited dramatic statistics to prove it. This was something James Pettit knew more than a hundred years earlier.

The Ottawa Tent Colony opened in 1904 on the south side of the Illinois River. After a few years, 9 x 10-foot wooden huts, sometimes called cottages, replaced the canvas tents. There are many testimonials to the success of this treatment.

A sanitarium for indoor treatment was built in Ottawa in 1918. By 1921, the tents and huts were no longer used by patients. Many of them remained on the property until the 1950s.

The Club House, where patients took their meals.

2

THE OTTAWA TENT COLONY

It was in January 1904 when the Illinois State Medical Society announced a war to fight tuberculosis would be a primary focus of its annual meeting as part of the nationwide and worldwide crusade that was becoming a phenomenon.

The Illinois State Medical Society was organized on June 4, 1850, when twenty-nine delegates from local medical groups gathered in the Old State House in Springfield. The meeting, called by the Ottawa Medico Chirurgical Society and co-sponsored by the Aesculapian Society of the Wabash Valley, was held to reorganize the Medical Society of Illinois, a loosely organized and informal society founded in 1840. Dr. Pettit told the *Chicago Tribune* in January 1904:

> There is no fact in medicine better established than that tuberculosis can be prevented and cured. Notwithstanding this, about one-seventh of deaths from all causes are due to this disease. The mortality should not be greater than that from smallpox, diphtheria or scarlet fever, and would not were the means at our command generally understood. We propose to spread this knowledge, for the solution of the problem depends upon the education of the public.

Pettit said the statistics presented at the convention would be confined to Illinois "and will include mortality, influence of climate, topography, occupation and residence."

In a letter to the *Chicago Tribune* published on January 25, 1904, Dr. Pettit thanked the newspaper for its coverage of the medical society's goals in waging war on tuberculosis. He wrote:

> We are not seeking to exploit some new and untried medical fad, but agencies which are at the command of almost every one. These are mainly the scientific application of fresh air, nutritious food, rest and exercise. But what is more important, we want to teach the public how to prevent the disease and thus save from 6,000 to 7,000 valuable lives in this state each year. This claim may seem extravagant, but if given an opportunity we will prove it.
>
> This movement is purely philanthropic. Our recommendations, if followed, will not in any way inure to the advantage of the medical profession, but solely to those who

have the disease, and that increasing army who are each year becoming infected for lack of simple, yet effective, precautions.

At the Illinois State Medical Society's annual meeting in Bloomington in May 1904, the 500 physicians chose Dr. Pettit to head the war on consumption. The society pledged to raise $10,000 for the treatment of tubercular patients.

The Ottawa Tent Colony opened on July 1, 1904, under the auspices of the Illinois State Medical Society to demonstrate the curability of tuberculosis in Illinois. It had just two patients on opening day. The medical directors were Drs. J. W. Pettit and E. H. Butterfield.

The tent colony was on a 120-foot-high bluff overlooking a ravine leading to the Illinois River. Ellis Park, a popular entertainment venue with a theatre, a pavilion, and other amenities, and which was served by a streetcar line, was nearby. The park ceased not long after the colony opened.

The Ottawa Tent Colony accepted only early cases of tuberculosis. Its opening capacity was for sixty patients. It was a private facility, not a state institution, and rates ranged from $18 to $30 per week.

The patients stayed outdoors in canvas tents. After a short while, wooden huts with canvas flaps became available. The tents and the huts were equipped with electricity for lights and for electric blankets, which kept their bodies warm even if the blankets sometimes had an inch of snow on them.

Before the permanent administration building (called the Club House) opened in December 1905, where patients could take their meals indoors, a large tent was erected for a dining room. Cooking was done outside, and food was purchased from local farmers.

The building was used as a club house, administration offices, living quarters for nurses and other staff, a lounging room, kitchen, and dining room. Paper plates and cups were used, then burned.

O.T.C. is the logo for the Ottawa Tent Colony.

Right: Ellis Park was a popular entertainment venue, located next to the tent colony. This newspaper advertisement is from July 1, 1904, the same day the Tent Colony opened.

Below: Picture postcards showing the Ottawa Tent Colony property.

811 Ottawa Tent Colony, North View, showing Illinois River, Ottawa in the distance, Ottawa, Ills.

Illinois River, Looking West from Tent Colony.   Ottawa, Ill., in Distance

Ottawa Tent Colony Club House and Grounds, Ottawa, Ill.

CLUB HOUSE OF THE OTTAWA TENT COLONY.

The dining room was open on two sides to let the fresh air blow through. There were a bathhouse and a kiosk for relaxing.

The emphasis stressed nutrition, with three meals a day and two additional lunches. The patients were served five or six quarts of milk, six to twelve raw eggs, meat, and more. Instead of wasting away, as tuberculosis did to its victims, most patients gained weight. A news item in the *Ottawa Free Trader* on October 25, 1907, noted:

> Miss Christine Liska of Milwaukee, who has been a patient at the Ottawa Tent Colony and is known as the sunshine of the colony, returned home Saturday, cured after a year's treatment for tuberculosis. Miss Liska gained 55 pounds.

Eating all that food was not a chore because being outdoors worked up a healthy appetite. This addressed the two main concerns—fresh air to heal the lungs and plenty of food to avoid losing weight.

A story in the *Ottawa Daily Republican* on the day before the colony opened reported:

> The discipline of the camp will necessarily be severe. A restoration to health requires an absolute certain line of conduct and diet. Health will be regained by the observation of natural laws, administered on scientific principles. One of the first things to be observed will be rigid enforcement of a rule against spitting. Tuberculosis is contagious only from the evaporation of the saliva and the dried particles floating in the air. The too-common practice of spitting everywhere will be barred absolutely on the grounds, the patients being provided with the necessary cups and receptacles for the care of the sputum. Regularity of living will be another rule which will be rigidly enforced. Breakfast will be served promptly at 7 o'clock; at 10 o'clock the patient will be given a quantity of milk and raw egg; at noon, dinner will be served and will consist of beefsteak and roast beef, with vegetables and fruits of all kinds and ice cream dessert, but no pastry or confections. Again in the afternoon, another ration of milk and an egg will be taken. Supper will be served at the usual hour and of an ordinary character, and at 9 o'clock more milk and egg and then to bed. Six meals a day, fresh air, rest, quiet recreation and sleep make up the prescription for the patient, whose hope of life hangs by a thread. Yet so thoroughly has this prescription been tried that the results are not problematical, they are certain.

The public was not allowed on the grounds unless invited, to keep anyone who was curious away.

The *Chicago Chronicle* printed a full-page story about the tent colony on Sunday, July 24, 1904. The colony had been open just a few weeks, but the visiting reporter was impressed with what he saw. Numerous pictures were included with the story. The reporter wrote what had been mentioned before, about the fresh air and nutrition, the climate not making a difference, and the beautiful scenery on the bluff overlooking the river. There were just four patients at that time, but it was obvious they had improved greatly in their few weeks there. It was noted that the grounds had 100 signs painted on wood boards and nailed to trees, all different and all warning against spitting.

Dr. Pettit hoped the state would take his fledgling tent colony and make it a state institution. He led a delegation of twenty physicians in 1905 to lobby

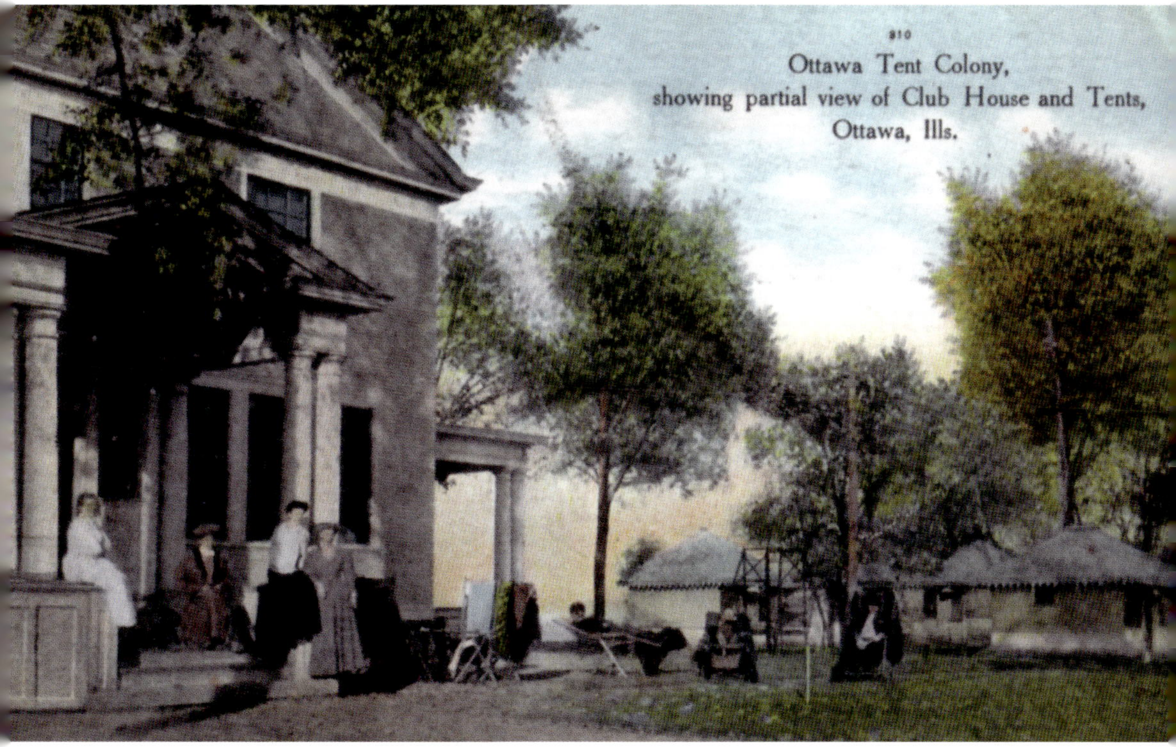

Ottawa Tent Colony,
showing partial view of Club House and Tents,
Ottawa, Ills.

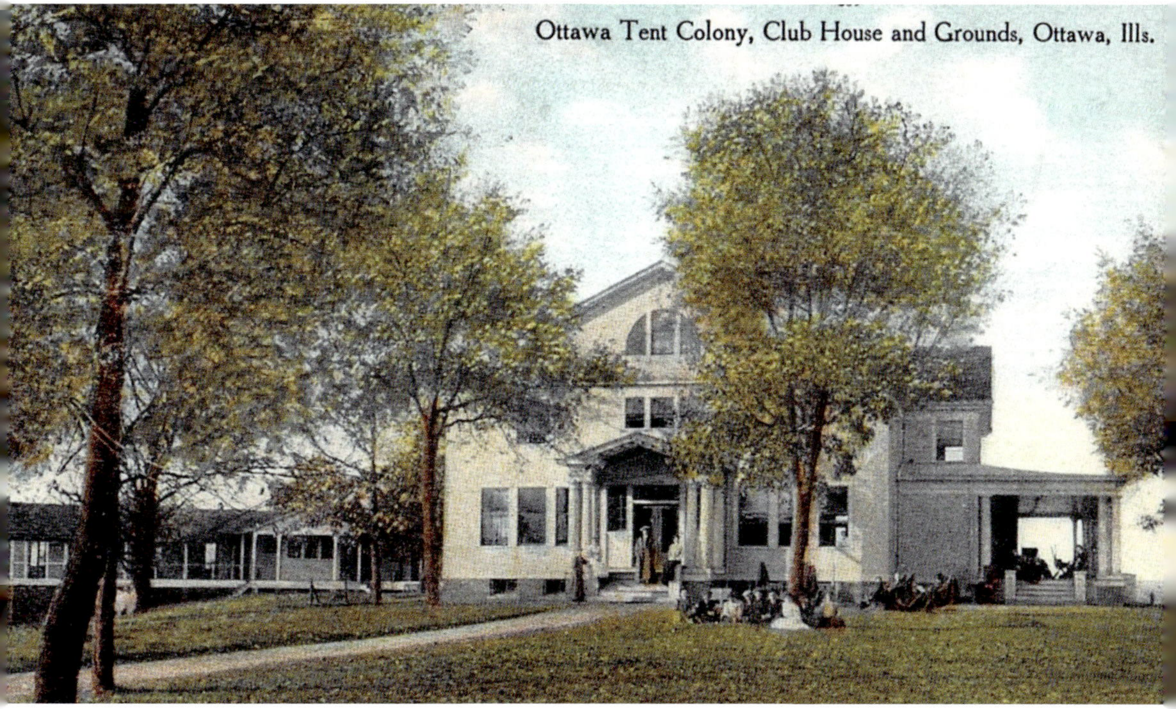

Ottawa Tent Colony, Club House and Grounds, Ottawa, Ills.

the state legislature to appropriate $200,000 to establish a state tuberculosis sanitarium. It was to be located near Ottawa and use methods applied by the tent colony. Pettit told the legislators that 8,000 people in Illinois die each year from tuberculosis. He said patients in other states that use his plan have a 76 percent recovery rate. Dr. George Webster told the legislators that more people in Illinois die from consumption than from typhoid fever, smallpox, scarlet fever, diphtheria, and erysipelas combined.

The state senate passed a bill appropriating $100,000 to build a T.B. sanitarium. The house cut the figure to $25,000. Dr. Pettit and the state medical society knew that was not enough so they told Governor Charles Deneen to veto the bill. It was decided that the T.B. sanitarium would be a private institution instead of a state institution.

Dr. Pettit said at the dedication of the administration building in December 1905:

We had hoped to make this a charitable institution, but we could not get the legislature to give us enough money. Credit for success is due to the patients who lived here through the last winter. We tell patients the truth. We don't cure them. We teach them how to live and send them home to cure themselves. We have demonstrated that tents can be used in cold climates and we have reached hundreds of thousands of sufferers.

The first thing proved is that climate has no special significance. The second, that in the matter of equipment the treatment can be had much more cheaply than ever before. And the third, that food is of the utmost value. If anything, I would place this last thing first. We do not do anything here but eat and rest and breathe God's air. It costs $1.33 a day to feed each patient. They get the best we can buy.

Sanicula Springs on Ottawa's south side provided the mineral water for the tent colony.

It is the duty of the medical profession to arouse itself and send us patients in the early stages. Modern treatment must be recognized. It is only the early stages that have a good chance to get well. The fate of the patient is in his own hands. It is a matter of character, largely. In this, as in life, the weakling must go down.

The public needed to be educated in the simplicity of the treatment of tuberculosis—that was the subject of a meeting of the Chicago Medical Society on October 12, 1904. Papers on the subject were read by Drs. Pettit, Homer Thomas, and Charles Mix.

Dr. W. E. Quine, the new president of the Illinois State Medical Society, told a group at the Chicago Public Library on December 17, 1904, that storm doors and storm windows help spread tuberculosis because they prevent airflow and ventilation in a house. In 1905, Dr. William A. Evans of Chicago suggested that all large cities put up shacks and tents on vacant lots where poor people with T.B. could get the open-air treatment.

Samuel Hopkins Adams wrote an article in the January 1905 issue of *McClure's Magazine* on the nationwide crusade then being started to fight tuberculosis. He wrote that almost every country in the world had started organizations to combat the disease. Every major city in America had established clinics, too.

It was in harmony with this general movement of the most enlightened medical, scientific and humanitarian thought of the world that the Illinois society began its fight last spring.

One of the most practical things done by the Illinois State Society was the establishment near this city of a Tent Colony for the purpose of demonstrating the curability of this disease in this climate. Still in the experimental form, its work is being watched by the medical fraternity everywhere. And results have been accomplished that have carried a message of hope and encouragement to stricken ones all over the country.

Adams' lengthy article covered the facts of the disease and that social conditions, such as dirty, overcrowded slums, were a large factor in the spread:

First, it is the chief cause of death around the world. Second, it is the greatest of all drains on a nation's resources since it disables from one-quarter to one-third of the population at the productive age. Third, that the one serious source of infection is from man to man by the sputum expectorated or coughed up. Fourth, that although communicable it is not contagious; therefore the careful and intelligent consumptive is never a public peril. And finally, that it is always curable.

Tuberculosis is a house infection. We don't pick it up on the street as we may pneumonia or smallpox. We never inherit it. Seldom is it contracted from diseased milk or meat. Occasional contact with a consumptive endangers no one; the disease is not contagious in that sense. But every house in which an ignorant or careless consumptive has lived and coughed up the deadly bacilli; every close and foul-aired work room in which he has labored, becomes a peril to those who live or work with him or follow after him.

19

Summer scenes at the tent colony.

Winter scenes at the tent colony.

The writer of this postcard said, "The young lady standing near the tree between two young men is Mll. Claudie."

Adams cited Chicago as an example of what was being done wrong. There was no law requiring registration of tuberculosis cases, so the data on deaths was unreliable. When someone died, the board of health did not disinfect the premises. The city of Chicago spent very little on public health; in fact, Adams wrote:

> The municipality saves money and spends lives. In the face of this, Mayor Harrison's recent message notes approvingly that the city shows a smaller outlay per capita on its health department than any other large city in the country. This is as if a man should contemplate himself admiringly in a looking glass and say, "I bet I buy less soap than any fellow in town!" The City Hall is what one might expect from the mayor's boast. It feels like a cave, smells like a cellar and presents to the visitor's eye, in its passageways, an underfoot prospect of uncleanliness comparable only to a slum bar room floor at midnight.

Adams agreed with the open air treatment that was prevalent at the time:

> The sufferer doesn't need to go to Arizona or California. Climate, while it may be an aid in some cases, has much less influence on tuberculosis, except in the later stages, than is generally supposed. Fresh air is the thing, and the air of Maine is about as good for this purpose in a majority of cases as the air of New Mexico, while the atmosphere of Illinois can fairly challenge comparison with either of them. Last winter was a pretty severe one in the Adirondacks; the mercury spent more time below the zero mark than above it, with an occasional excursion as far as fifty degrees; yet consumptives lived and slept in open shacks right through the season, getting hungry and fat and strong on it.

This is the menu booklet for the 1916 Christmas dinner in the tent colony dining hall. Served were blue points on a half shell; salted pecans; celery hearts; hothouse radishes; consumme printanier with paste croutons; broiled shad with bacon, Maitre d'Hotel; Saratoga chips, roast young turkey with chestnut dressing; currant jelly; fried potatoes Parisienne; whipped hubbard squash; frozen Christmas punch; pineapple and marshmallow salad; preserved cranberry sundae (a 1915 fruit cake); individual Christmas basket; and café noir.

Right: A view of the Club House from the 1940s.

Below: The dining room in the Club House.

Harriet Taylor told the Streator Club on November 16, 1905, about her visit to the Ottawa Tent Colony. The *Ottawa Free Trader* reported:

> Mrs. Taylor spoke enthusiastically of Dr. Pettit and his grand work, saying that he is just the one to take charge of such an undertaking to make it a success. She feels sure that it would pay everyone in Streator to visit the Ottawa Tent Colony when in that city, where they may view the most beautiful scenery and a fine new club house built for health and comfort. The great trouble the doctor has found, she said, is that so many patients have waited until the disease is so far advanced that most of them can only be helped. Out of two hundred patients, he has had eleven incipient cases. These have all improved rapidly and he belives permanently. Therefore, to ward off this great calamity of disease, early diagnosis and no spitting on our streets is urged.

During the week of December 4, 1905, several banquets were held in the new Club House building for newspaper editors, prominent citizens, and contractors and workmen. The *Chicago Record-Herald* devoted a lot of space to the colony and the new building that week. Dr. Pettit again stressed that T.B. patients can recover no matter the climate:

> Living in a tent in southern California or New Mexico may be more comfortable than in Illinois, but the patient will not recover faster. A consumptive cured in the Illinois climate should continue to dwell in that climate, and one cured in California should continue in that climate. In short, the medical world can say to former consumptives, "Stay in the climate in which you were cured."

Dr. Pettit added that early treatment is a key to a patient being cured. He also said that although the Ottawa Tent Colony was a private facility, he would continue to work to establish a state institution.

The bacteriological laboratory at the tent colony.

There was a small disagreement at the state medical society's meeting in December 1904. Dr. Arnold Klebs wanted the committees to be comprised of physicians. Dr. Pettit insisted that laymen be included, so they were.

Dr. Frank Billings told an audience at the tent colony in December 1905 that "you can get pure air in Chicago" and they needed to be told that fact. He said city people think they have to go to the country for fresh air, while at home "they pull down the windows, shut the doors, stuff up the cracks that might admit a little fresh air, and if the weather is a trifle cold, they hurry when necessity takes them out of doors for a few minutes."

> The consequences are inevitable. The constant breathing of vitiated air conduces to tuberculosis and makes the slightest draft a matter of serious danger. The general health is so debilitated that even a slight attack of illness is a menace. The breakdowns due to a lack of pure air are without number.

Dr. Billings said many areas in Chicago are "befouled by smoke and stenches," but there is fresh air in the city, especially from lake breezes.

Editors from across LaSalle County met for an association meeting at the courthouse on December 5, 1905, and then adjourned for a luncheon at the tent colony. "Milk, eggs meat, particularly beef and mutton, probably rare, potatoes, rice, beans, peas, fruits as appetizers, nuts and more milk"—this was the regimen for tuberculosis patients given in a *Chicago Tribune* article on April 5, 1904. Fresh air and good food was a "sure cure" for all but those in the most advanced stages, Pettit said. His remarks were given at a conference of the Chicago Tuberculosis Exhibition in the Chicago Public Library. The topic was "The Outdoor Treatment of Tuberculosis." Harriet Fulmer, head of the Visiting Nurses Association, told the large crowd:

> While the tubercular conditions among the poor, due to poor food and bad housing, are still grave, the visiting nurses can detect an improvement. Where a few years ago there were ten windows open to the fresh air, there are 100 now, and the need of cleanliness, good food and sunlight gradually is being realized.

The *Ottawa Free Trader* printed the following on December 13, 1905:

> Among the hundreds of physicians and prominent lay people from all parts of Illinois who visited the Ottawa Tent Colony on Tuesday were Dr. J. C. Roberts, Dr. G. Zeller, Mrs. Clara P. Bourland, Julia Keyes and Secretary Casper of the Associated Charities from Peoria. On their return to that city in the evening, the Star represents them as "marveling much at what they had seen and heard and were unanimous in the opinion that the outdoor treatment for consumption was simple and effective."
>
> They found there, adds the Star in more detail, between 50 and 60 tents occupied by tuberculosis patients and a large building which has recently been erected at a cost of $20,000 where meals are served and the patients take their exercise in the inclement weather. At night they sleep in tents which are exposed to the elements, no matter what their conditions. Their beds are warm and their coverings comfortable, but after they have retired, attendants open the ends of the tents and allow the breezes to blow through. Thus, they sleep while the thermometer may be many degrees below zero and actually derive benefit from it.

Since the tent colony was started two years ago, over 200 tuberculosis patients have been cured with the open air treatment and proper diet. They eat five meals a day of simple but nourishing food and the open air at night does the rest. The Ottawa Tent Colony was started as a private enterprise by Dr. Pettit two years ago on a small scale, but he was enabled to demonstrate what could be done by the treatment and now there are 45 patients there.

In a brief talk, Dr. Zeller stated that while over 200 cures had been effected, the value of the treatment has been demonstrated to thousands and he regretted that it was beyond the province of the great state of Illinois to supervise the system but had left it to private enterprise to bring it to a successful issue.

The Ottawa Tent Colony's $15,000 Club House and administration building was officially dedicated on December 12, 1905. Approximately 1,000 members of the medical profession from across the state attended.

The *Ottawa Free Trader* observed:

It has shown, beyond all room for disputation, that the dietetic treatment of pulmonary tuberculosis is no longer an experiment but an assured success. This, it is justly claimed, makes it the duty of every physician to familiarize himself with the methods of treatment and equipment here pursued as necessary to their successful application. No amount of reading will convey as much information as can be obtained by a short visit to a tuberculosis sanitarium. It is surely at least a matter of interest to see how comfortably patients can be housed in tents in winter time. To this end, invitations were generally sent to the medical profession of this state as well as to prominent laymen to meet at the Ottawa Tent Colony tomorrow.

The *Chicago Tribune* observed there were sixty army-style tents on the bluff over the Illinois River:

Men and women in all stages of consumption live in these tents day and night, in winter and summer. From the few tents put up in July of last year, the colony has grown until the institution includes water works, bath house and an infirmary, beside the new building. Dr. J.W. Pettit claims that 90 percent of incipient cases can be cured by his treatment.

The guests were served a buffet luncheon in the Club House, the same meals as the patients received, with coffee added.

The main speeches were given by Drs. Frank Billings of Chicago and Pettit. Dr. George Zeller, superintendent of the Illinois State Hospital for the Incurable Insane at Bartonville, made a few remarks. Also speaking were Harriet Fulmer, director of the Visiting Nurses Association; Ottawa banker and former mayor A. F. Schoth; labor leader Duncan McDougall; Isaac Mitchell, principal of Lincoln School; Will O. Clark, proprietor of the Clifton Hotel; and Fred Sapp, editor of the *Ottawa Republican-Times*. The Gualano Brothers band played musical selections.

The *Chicago Tribune* reported the speech given by Dr. Frank Billings: "When you come to think that one person in every seven dies of tuberculosis, it is a subject that cannot be talked of too much." Dr. Billings said tuberculosis is not inherited, as some people believed. People who can go to a hot, dry climate should go if they can afford it, "but thousands are sent away to Mexico and California who die alone, living in poor rooms and getting poor food." Billings continued:

> You can get pure air in Chicago. I had a patient who was cured by building a sleeping cage out of a second story window and filtering the air through cheesecloth. Physicians are not careful enough in diagnosing the disease, especially among poor people. If recognized in time, a great many cases can be cured absolutely.
>
> What has been accomplished here by Dr. Pettit is a revelation. It does not exist anywhere else in the world. It has proved that a tent is more desirable than a habitation of wood or stone. Much has been accomplished for humanity here. And it is a treatment that is within the reach of all.

A historical account of the dedication in the *Journal of the American Medical Association* said:

> Dr. Frank Billings made the principal address, in which he told in plain terms the latest decision of science regarding the treatment of tuberculosis, and eulogized Dr. James W. Pettit for his work in the cause of humanity. Two years ago, the state medical society authorized Dr. Pettit to make this experiment at his own expense. In closing, Dr. Billings said the greatest work that Dr. Pettit has done has been to humanity in general by setting a great living example of right living to all the world, "and these things I say not in flattery but in envy of the great work that Dr. Pettit has been permitted to do at a great sacrifice to himself." Dr. Zeller praised Dr. Pettit for reducing the per capita cost for the care of tuberculosis patients and stated that the modern method of treating tuberculosis was being followed as far as possible in the state institutions.

Want ads in the *Chicago Tribune* tell us what workers at the Ottawa Tent Colony were paid. In 1917, undergraduate nurses were offered $30 per month, including room and board. The same offer was made from 1914 to 1916 for a "man, neat appearing, young, to wait on tables and make himself generally useful." A head waitress in the dining room was offered $35 a month in 1909, plus room and board. A maid was offered $22.50 per month with room and board in 1916. A man to fire the boilers and do other work was advertised at $35 a month in 1917. Female housekeepers were paid $35 a month in 1919. Nurses were paid $40 a month plus room and board in 1920. Young men needed to work in the dining room were paid $45 a month in 1920. A female nursing attendant, with no experience necessary, was offered $50 a month and room and board in 1920.

The second annual conference of the Chicago Tuberculosis Institute was held on April 4, 1906. The subject was "Outdoor Treatment of Tuberculosis." Dr. Pettit was a speaker, along with Drs. H.B. Favill, William Evans, Theodore Sachs, and Ethan Gray, along with Harriett Fulmer, superintendent of the Visiting Nurses Association.

The National Association for the Study and Prevention of Tuberculosis held a seminar in Manistee, Michigan, in July 1906. Dr. Pettit spoke on "The Modern Treatment of Tuberculosis." He again stressed that tuberculosis patients did not need to be sent to warmer climates in the west, that any climate is as good as any other. He also stressed the importance of nutrition. The *Chicago Tribune* of November 11, 1906, included:

The charge that at the Ottawa colony patients were "stuffed" with food was refuted by Dr. Pettit, who showed that the increased combustion caused by the open air treatment and the wasting nature of the disease demanded greater quantities of food. It was shown that home treatment was apt to be deficient in respect to the nutrition of patients.

The pursuit of the simple life, with plenty of fresh air, abundant and nutritious diet, exercise and rest, is the correct and practical treatment for pulmonary tuberculosis, the great white plague, according to James W. Pettit of the Ottawa Tuberculosis Tent Colony. Dr. Pettit lectured on the "Tent Treatment of Tuberculosis," the second of the Chicago Medical Society's free lectures in the Public Library Building last night. The Ottawa colony, established several years ago by the Illinois State Medical Society to solve the problem of treating consumption, has proved that the disease can be cured by the open air method. Climate conditions are unimportant and drugs are useless. Life under simple and primitive conditions is the most practical and economic system. Dr. Pettit illustrated his discourse with illustrations of the Ottawa camp and patients who have been cured by the open air method.

Alice Nelson died of tuberculosis at the age of twenty-five on September 21, 1907. She was a graduate of Pleasant View Luther College on Ottawa's south side. She had been ill for three years.

Mississippi Governor J. A. Vardaman and members of the state board of health toured the tent colony and had dinner there on October 31, 1907. Vardaman was looking for ideas in establishing a TB clinic in his state. The *Ottawa Free Trader* noted, "Governor Vardaman is a typical southerner, from the top of his soft black hat, which covers a flow of long hair, to the bottom of his well-polished shoes."

In May 1908, it was reported that the tent colony had a capacity of sixty patients, with Drs. Pettit and Butterfield as medical directors. Rates were $18 to $30 per week.

A special homecoming celebration was held at the tent colony in June 1908. It was organized by Dr. Butterfield to bring back former patients who were cured. Invitations were mailed, and about thirty people showed up; letters were received by many other former patients. The newspaper reported:

The meeting of old time patients at the Colony was like that of so many college students come back to alma mater. Only in this case, the rah-rahs meant more. A fine dinner was served [to] the visitors of the Colony. While it was going on, Prof. Joseph Reardon entertained as only he knows how to do. Afterwards, Dr. J.W. Pettit presided as toastmaster. Then, many of the former patients spoke words of praise and appreciation for an institution that is making Ottawa noted throughout all this great central west. Altogether it was a great and encouraging gathering.

A group of teachers from the LaSalle County Teachers Association toured the tent colony in September 1908. Dr. Pettit addressed their institute meeting and offered an invitation. They took him up on it. Harley Pettit gave them the tour and explained the functions. The group looked at the tents, the hospital annex, the kitchen, offices, and laboratory before being served supper in the dining hall. The teachers' newsletter said:

The location is certainly ideal, being situated on a high bluff overlooking the beautiful Illinois River, and having a natural wood for shade. The surroundings of the patient are as inspiring and pleasant as could be found anywhere. To inform the children how to avoid disease is a fundamental duty of the modern teacher. Undoubtedly, the trip to the Tent Colony will result in much good to many pupils of LaSalle County.

The Illinois State Conference on Charities held a program in Rock Island in October 1908 that featured an address by Dr. Pettit on the difference between local and state sanitariums. He criticized politicians who built expensive institutions without consulting people in the medical field, calling it "the footballs of partisan politics and the monuments of vanity."

Dr. Alexander Wilson of Chicago advocated a program that would educate the public on the prevention of tuberculosis. Dr. Adolf Meyer spoke about the treatment of insane people. Dr. Meyer, a former pathologist at the Eastern Illinois Hospital for the Insane in Kankakee, Illinois, was the chairman of the psychiatry department at Johns Hopkins University in Baltimore and was considered one of the world's leading authorities on psychiatry.

Dr. Pettit addressed a large crown at the Congregational Church in Ottawa on Sunday, January 3, 1909, on what was being done to fight tuberculosis. His talk included showing pictures on lantern slides.

Dr. E. H. Butterfield sold his interest in the tent colony in May 1909 to Harley Pettit, the superintendent and the son of Dr. Pettit. Dr. Butterfield took a position as the physician at the nearby Buffalo Rock Tent Villa. He was the son of Orcott Butterfield, a hardware store owner in Ottawa.

Dr. Pettit addressed the relationship between the medical community and the press at the annual meeting of the Illinois State Medical Society in May 1909. "No agency is more potent for good or evil than the secular press," Pettit said. He also said truthful advertising was all right, but not ads for fraudulent medicines.

Dr. Pettit addressed the cost of tuberculosis at a conference on state charities in October 1909. He said the cost of tuberculosis could be seen in four aspects—the cost in lives, disability, unhappiness, and money. Dr. Pettit remarked:

The cost in lives in the United States is estimated at 150,000 annually, which is more than the combined deaths from typhoid fever, scarlet fever, smallpox, diphtheria, cancer, diabetes, appendicitis and meningitis. If a forecast of the deaths of the future is computed on the basis of those now living, and the death rate in the past, it is safe to say that five million persons now living in the United States are doomed to die of tuberculosis unless present conditions are changed. Of this number, about 9,000 died annually in the state of Illinois.

The coverage in the *Chicago Tribune* continued:

Since it is easier to engage public attention upon the financial cost of a public burden than upon loss of life or suffering, I may be pardoned for dwelling upon this particular feature, which while not be the most important, is the one upon which we can more easily command public attention.

Dr. Pettit said the cost was not just in medical treatment; it also was the loss of earnings and the cost of burials. He cited a study by Professor Irving Fisher that the annual loss in the United States to tuberculosis was $1.1 billion. Dr. Pettit said:

> There is more than enough money spent in one way and another which if properly directed would cure the curable and make harmless the incurable. Almost every consumptive who dies has actually cost more in dollars and cents than it would have taken to cure him if the disease had been discovered and treated in time.

He also said it was "cruel and inhuman" to send dying patients to other climates for treatment:

> Fully 7,180 persons hopelessly diseased with tuberculosis annually go to die in the states of California, Arizona, New Mexico, Texas and Colorado, most of them by order of their physicians, and many of them too poor to afford the trip. Consumption can be cured, or arrested, in any section of the United States, and the percentage of cures in the east and west is nearly the same. Any physicians, therefore, who sends a person to the southwest without sufficient funds, or in an advanced or dying stage of the disease is guilty of cruelty to his patient.

Efforts were being made, he added, to stop this and instead build small local hospitals in every city: "Attempts also are being made in southern California and in Texas to exclude indigent consumptives or send them back to the east."

J. N. Young of rural Henry, Illinois, placed a classified ad in the *Chicago Tribune* in December 1909: "Personal. Will some person of means help effect the tuberculosis cure of a deserving Christian mother? Address CO22, Tribune." The Youngs had five children. Mrs. Young was a patient in the Ottawa Tent Colony. She had contracted tuberculosis two years earlier, and their house, which was paid for, had to be mortgaged to pay for Mrs. Young's treatment. Local doctors could not help, so she was taken to the tent colony. His ad was a plea for funds to help his wife.

A field trip for eight male and female patients caused injuries when the bus overturned on Van Buren Street on August 26, 1909. A pole pulled loose from a front axle and the passengers were thrown out before the horses could be stopped.

Ella Lindsay came to Ottawa from Peoria in mid-August 1909 to work in the dining room of the tent colony. She drowned in the Illinois River while swimming on August 25, at the south ferry landing leading to Bull's Island, after she had finished work. Her body was found down the river in Peru.

Henry Thomas Benton, a nationally famous journalist, included a colorful mention of the tent colony in one of his dispatches about his travels in October 1909:

> Out where the interurban railway track crosses Wolf Creek is a camp that was located for one who has been good to all and who now suffers from the affliction called the white plague. In that camp is Clinky Logan and for a companion he has with him Dago Eagle. The way they live and the way they are going to live through the coming winter shows how strong are the powers of fraternal friendship.
>
> Logan is an apprentice bottle blower and Eagle is a baker by trade, for his father was one and he taught his art to his boy. Not to enter into the medical part of the white

plague, it is enough to be said that Logan has it and it is the intention of his friends to see if he can get over it.

Where he is at camp he is under as good a tent as was ever made, for it is one that was brought and used one season by the Illinois Valley Fishing Club. Logan has with him in the tent, a kerosene stove to cook with and also to make the frigid air of these nights agreeable. He has lights, he has fire, he has the regard of his friends, and best of all, he has human sympathy.

His tent is being fixed this afternoon by a company of his fellow workers and when the cold, chilly winds of December blow he will be as snug as a kitten by the kitchen fire. Clinky may live and he may not. But what Dr. Pettit of Ottawa teaches in a group upon the hills south of Ottawa, the members of the Green Glass Bottle Union are doing for Logan. They do so with a kind belief that he will live if he has fresh air and a diet that any physician would approve. A party of Logan's friends went out to see him this afternoon on the interurban and nothing will be left undone to see that with him all is well.

The Western Review, a Chicago magazine, recognized Dr. Pettit in its December 1910 issue. The article cited Dr. Edward Trudeau as a pioneer of the outdoor treatment at his Adirondack sanitarium and said Dr. Pettit was the first physician in the Midwest to recognize this treatment and set up an outdoor facility:

The writer has visited this colony occasionally and has kept fairly well informed concerning the work done there since its inception. The work being done here is inspiring and encouraging and of incalculable benefit not only to the patient undergoing treatment but to the present and more especially to the succeeding generations.

The anti-tubercular movement which is now sweeping over the civilized world has already accomplished much good by creating an interest in a subject which has hitherto been treated most indifferently. A movement of this kind, in order to be of lasting benefit to humanity, must have a more substantial basis than the mere education of the public through the written or spoken advice. The hygienic-dietetic cure of tuberculosis is no longer a theory but an established fact. It has also been demonstrated that tuberculosis can be cured in any climate if the patient will only accept the conditions which the treatment imposes. No method of treatment of pulmonary tuberculosis has given such universally satisfactory results as that adopted in sanatoria devoted to the treatment of this disease.

The writer said that between 1904 and 1910, there had been 1,200 patients admitted in all stages of the disease with a 35 to 40 percent rate of cure, with perhaps a 85–90 percent cure rate for those in the early stages.

Dr. Pettit wrote a lengthy article for the *Illinois Medical Journal* in February 1911, reporting on the progress that had been made in the six and a half years of the tent colony's operation. He wrote that there had been 1,100 cases admitted to the Ottawa Tent Colony at that time. He put the cure rate into three classes based on the initial stages of the patients—90 percent cured for those in the early stage, 50 percent of those in an advanced stage, and 1–2 percent cured in the far advanced stage. Here is an excerpt from his article:

In addition to the usual hygienic-dietetic methods, we have administered tuberculin consistently and conscientiously during the past four years. For some time we have followed the clinical methods of Trudeau, Koch and Denys. The latter method is much less laborious, and as we believe, more scientific.

At the inception of this work, we were confronted by that hopeless apathy which had hitherto characterized our attitude toward tuberculosis; also a very general skepticism both in and out of the profession as to its curability in this climate. I think it is fair to assume that public and professional opinion has changed on this question to such an extent that now it is generally conceded that tuberculosis can be cured practically anywhere.

The essentials, so far as climate is concerned, being simply fresh air. This change of sentiment has been brought about in a comparatively short time when we take into account the magnitude of the problem, and the time-honored opinions and convictions which had to be overcome. Now that former prejudices, to a great extent, have been over come, we are entering upon an era of constructive work.

One of the greatest obstacles to successful treatment is the fact that not only the Ottawa Tent Colony, but other institutions, are being compelled to admit many patients that are well beyond the incipient stage. The difficulty, from the standpoint of the sanatorium physician, is that he is compelled to accept a majority of advanced and far-advanced cases which makes successful treatment doubtful even under the most favorable circumstances.

The successful treatment of tuberculosis as applied to a large majority of cases can only be accomplished by removing the patient from the temptations of the home, business, and friends, and the sooner this fact is recognized, the better will be our results. The usual argument in favor of home treatment is that because of the meager sanatorium accommodations, and the poverty of the majority of the sufferers, treatment must be carried out in the home or not at all. This is unfortunately true, and the argument would have more weight were it not that the plan is so uniformly unsuccessful. A more rational plan is to meet the situation squarely by providing sanatoria for all. The failure in the home has a decided tendency to discredit the treatment and emphasizes the necessity for greater sanatorium facilities. It is unfortunately true that the modern treatment of tuberculosis can only be successfully applied with any great degree of success in sanatoria, hence we should devote our energy to making institutional provision for all classes of cases.

In addition to the temptations inseparable from the home, another important factor in favor of the sanatorium is the effect of the example of other patients, and that *esprit de corps* which is developed among them, and both of which are powerful factors. However intelligent and purposeful, determined and obedient, the patient may be, he will advance much more rapidly in having the stimulus which is obtained from life with other patients who like himself are seeking to recover health and whose example will constantly urge him on toward the proper life. All who have handled many patients together know the great assistance which is obtained from the rivalry between them in taking the cure faithfully; the brace to the weak from the example of others; the encouragement to the discouraged that is incurred by the good results in other cases; the mutual assistance and the good spirits which animate them all.

Dr. Pettit wrote that a major obstacle for many patients is being able to pay for the treatment:

However, there is no escape from the financial burden of tuberculosis. It must be borne in either one of two ways—in care or cure. The cure is the least expensive. It is a lamentable fact that there is as much money being spent in the care of tuberculosis as would be necessary to cure those afflicted if it were spent at the right time and in the right way.

He estimated the cost of a cure at \$300–500:

It will usually cost considerably more than this to care for him during the months, or even years of his illness, to say nothing of the loss of his own time and that of other members of the family who are compelled to give up their employment in order to care for the invalid. Hence, inasmuch as the expense must be borne and there is no escape from it, where the patient may fairly be regarded as curable, he and his friends should be made to understand that a cure is usually less expensive than care. The resourceful physician can often be of greater service by his advice in this particular than in any other way.

Dr. Pettit explained how a lack of money sometimes prevents a successful outcome:

The word "cure" is often loosely used, thus giving rise to wrong impressions. Inasmuch as the active symptoms are more largely due to a mixed infection than to the tubercle bacilli, and this disappears under favorable conditions rather promptly, the patient and even his physician is apt to overrate the permanency of the results, and underrate the difficulties of completing a real cure. When we have overcome the mixed infection, the patient may have the appearance of health, but as a matter of fact he is not well, but simply placed in a condition where the cure of his tuberculosis is only fairly begun. This is a fact that is not appreciated, and every such failure is unjustly charged up to the defects of sanatorium methods.

While it is true that tuberculosis is easily cured, it is also true that it is never quickly cured, and the more fully this fact is appreciated, the better it will be for the reputation of the physician and the welfare of his patient. In the treatment of tuberculosis, like every other enterprise, the results are in proportion to the investment. Experience the world over proves that at least six months of sanatorium treatment is required to produce substantial results in the majority of cases, and that while a considerable number do succeed in a shorter residence, such a course is attended with much risk and many failures, which, but for the attempt to take a short cut, would have recovered. Unfortunately, many patients for financial reasons cannot remain long enough to effect a cure. It is fair to say that some treatment is better than no treatment, and not infrequently is successful, but such risks should not be assumed where possible to avoid them, and no patient, whatever his circumstances, should be left in ignorance of this fact.

The advantages of sanatorium treatment are that it affords the tuberculosis invalid the most practical and only systematic method of fighting his disease, and acquiring hygienic living in its prevention. The sanatorium is as essential as the general hospital, and for substantially the same reasons. Tuberculosis is a disease which heals slowly. This so-called recovery of health which follows change is often only apparent. It takes months to get well of incipient tuberculosis. If tuberculosis is ever to be mastered, it must come through early diagnosis and immediate, energetic, rational treatment. To attain this end, the earnest co-operation of the entire medical profession is required.

Dr. Pettit spoke to a medical convention in Aurora in March 1911, where he advocated a tent colony be established there. Dr. Pettit told the doctors that 5 percent of tubercular patients could be cured in this climate if they had proper treatment in the early stages of the disease. He said 30 percent of the patients sent to his institution who were in advanced stages were cured. The *Aurora Beacon* reported that the doctors were favorable to starting one. However, it was not done.

A large fire in the administration building at the tent colony on March 29, 1911, caused a small amount of damage. A spark from the chimney set the roof on fire.

The *Chicago Tribune* ran a story in April 1911 about a patient in Ottawa's tent colony who won a pool betting on the results of the mayoral election in Chicago. Peter Rolinski bet that Carter Harrison would win with a plurality of 17,132 votes—and that was the exact number. Rolinski told the newspaper that he dreamed he was walking down Madison Street when he saw a huge screen that said Harrison won by that number.

At the annual convention of the Illinois Medical Society in Aurora in May 1911, Dr. Pettit was elected chairman of the district.

It was announced in September 1911 that a $20,000 two-story building would be built east of the administration building. This was for the laboratory and other medical facilities. Ottawa architects White and Hanifen were given the contract. The plan was for two wings, on the east and west, each 30 by 70 feet. There would be porches and verandas, 10 feet wide, on the north and south sides. There would be sixteen suites for guests. Each room would have a private bathroom and a telephone.

Dr. William A. Evans, the former health commissioner for the city of Chicago and president of the Illinois State Association for the Prevention of Tuberculosis, came to LaSalle County in October 1911. He was campaigning in every county, urging officials to start clinics for T.B. sufferers. He spoke to audiences in Ottawa, Streator, LaSalle, and Mendota.

Dr. Evans was joined in the assembly at the Ottawa theatre by Dr. Pettit and several speakers. G. W. Perkins, president of the International Cigar Makers Union, cited statistics—every year, 4,000 people in Chicago and 176,000 people nationwide died of tuberculosis. Perkins said he insisted on cleaner conditions for his union shops, resulting in a reduction in deaths from T.B. from 51 percent to 24 percent in the previous twenty-three years. Dr. Evans said:

Consumption is a sin of society. It is produced by crowded, insanitary factories, poor houses and unventilated schools. There is no royal road of escape. We must learn the lesson and then live it. Consumption is semi-civilization. Chicago has 40-acre tracts where the population is 400 to the acre. Consumption uncontrolled must increase. Before it is wiped out there must be repentance. It will be driven out of the shops and tenements before it is driven out of the farm houses. All is not well in Ottawa. We should have a county sanitarium. But the authorities will not establish one except under the pressure of public opinion. That pressure should be exerted through this society and in other ways.

At this time, various groups were urging the county to purchase Buffalo Rock for the site of a T.B. sanitarium. That story is told in Chapter 6.

An article in the December 11, 1911, *Chicago Tribune* listed twenty T.B. sanitariums and dispensaries in the Chicago area; many were free for those who could not pay, and those were maintained by the Chicago Municipal Tuberculosis Sanitarium. Edward Sanitarium in Naperville and the Ottawa Tent Colony were included on the list.

A group of property owners pressured the Ottawa city commissioners in March 1912 to not go through with their plans to build a "detention hospital," or "pest house," for people quarantined with contagious diseases, near the pumping station. Approximately 100 people signed a petition. They did not want it near their property; they also objected because it would be close to the water supply and too close to a baseball diamond where small boys played. P. J. O'Gorman, Mike Dinneen, Edward Caton, Thomas Dixon, John Kilday, and Charles Funk headed the citizens' group.

Herbert Hoyt, a tubercular patient from Griggsville, committed suicide by shooting himself in the head on May 1, 1912, after becoming despondent. He had been a patient in Ottawa for six months; the doctors said he was on the road to recovery and was making plans to return home. He was forty-two and unmarried.

The National Association for the Study and Prevention of Tuberculosis reported in May 1912 that the death rate from T.B. in the United States had dropped in the previous decade from 197 to 100 per 100,000 people, while the death rate from all causes declined only half as much. It attributed the decline to increased sanitation and the campaign against T.B., particularly in bigger cities that had established T.B. clinics.

There was plenty of entertainment provided for the people in the tent colony. However, only a few were noted in the local newspaper. The Gualano family band from Ottawa played there often. They were very talented. After a concert in St. Petersburg, Florida, in December 1914, a newspaper critic there wrote:

Above: The new medical building, built in 1911–1912.

Left: Men sit in the medical building's pergola.

The violin numbers by Sig. Ettore Gualano has, since coming to St. Petersburg, been heard a number of times, but last night he was at his best. His playing carries with it the sense that the musician is putting himself into the music and his marvelous technique has been the subject of great comment.

A brief article in the January 21, 1908, *Ottawa Free Trader* said Alma and Ida Braun of Joliet entertained the patients with vocal, piano, and violin selections. On a regular basis, patients were treated to lectures on travel, medical information, and there was always chess, checkers, cards, and other games. There were outside activities like croquet, as long as it was not too strenuous.

The November 29, 1912, *Ottawa Free Trader* listed a long lineup scheduled for that night. There were eighteen acts on the bill that night and performers included Louis Ritzius, S. W. Raymond, Mr. Hathaway, Mr. Sanders, Mr. Jaeger, Mrs. Crowley, Mr. O'Toole, Mrs. Payne and Misses Brown, Coniff, Ritzius, Hutchings, Farrell, Price, Clark, Nertney, Cunningham, and Brunnick.

The Illinois State Medical Society met in Peoria for its annual convention in May 1908. Dr. J. W. Pettit represented LaSalle County as its main delegate. Dr. Pettit was elected president of the medical society at the convention. After the convention, Pettit went to Mudlavia, Indiana, to learn about the healing qualities of the mud baths and lithia waters there. Hotel Mudlavia was a grand hotel and spa near Kramer which was famous for its water, which was said to cure arthritis and other ailments. The hotel burned down in 1920. Often, such waters have healing powers because of radium in the water.

The Gualano family orchestra from Ottawa.

After Dr. Pettit ended his one-year term as president of the Illinois State Medical Society, he was chosen to represent the society at the national convention in Atlantic City.

The LaSalle County Board of Supervisors formed a committee in December 1912 to better care for poor T.B. patients. It had a thorough agenda—to have doctors, teachers, and other sources find the number and the location of people with T.B.; to determine which cases were in the early curable stages; to find which of those people could not pay for treatment; to determine what should be done for those in the later stages and to protect the public from them; and to prevent TB sufferers in the county poor farm from infecting others there.

Buffalo Rock, west of Ottawa, was the choice of many people as an ideal spot for a sanitarium. It was near the county poor farm and it already had medical buildings from a failed tent colony. The county board still balked at setting up a sanitarium. The county board statements added:

Nor are large and costly buildings needed for tuberculosis patients. Indeed, such buildings would of themselves be destructive of success. All that is needed in the way of housing is an open, shed-like structure to protect the patient from rains and snow, while leaving him the freest possible access to the open air. Such a structure, entirely suitable for the accommodation of ten patients and approved by the best scientific authorities, can be erected at an expense not exceeding $2,000 and one of these should be erected at the County Farm where the administration of it could be coupled with that of the farm, with but little extra expense, the farm supplying the food from its present kitchen, the farm physicians caring for the patients with the assistance of a head nurse, with perhaps at times one assistant. The items of expenses, in connection with such an installation, together with the cost of maintenance thereof, can be estimated with very close approximation before the meeting of the board in March.

Aiding the county board in this were Drs. J. W. Pettit and A. J. Roberts of Ottawa, Dr. Orr of LaSalle, Dr. D. S. Conley of Streator, Dr. C. A. Laffson of Serena, and Dr. E. K. Ayling of Tonica.

Dr. Pettit told the *Streator Daily Free Press* in a December 7, 1912, interview:

Where the county missed an opportunity was in not buying Buffalo Rock Tent Villa when it was offered for sale. The trouble seems to be the board wants to establish a tent colony on a cheap plan. If a colony is established, it will cost considerable money, but LaSalle County is rich and can easily afford to meet the expense.

The matter came up before the county board in December 1913. The board appropriated $3,200 in June for a T.B. sanitarium. It sought bids to build a facility, but the lowest cost from a contractor was $4,000. Nothing was done.

A gasoline stove in the kitchen of the administration building blew up on July 9, 1913. Quick work by the men put out the fire. The building was filled with smoke, and the extreme heat broke out many of the windows.

Governor Edward Dunne declared December 7, 1913, as "Tuberculosis Day," urging education on how to prevent the disease.

Dr. Roswell Pettit spoke to the Illinois State Association of Graduate Nurses in February 1914. He had just returned from Europe and studied the new vaccine therapy for preventing typhoid fever and for curing tuberculosis, scarlet fever, diphtheria, lockjaw, rheumatism, and infections.

The *Ottawa Old and New* book in 1914 said more than 1,500 patients from all over the nation had been treated in the previous decade.

In 1914, Dr. Pettit was listed as medical director; Dr. W. H. Jamison as resident physician; Dr. Roswell Pettit as bacteriologist and pathologist; H. U. Boudreau as bacteriologist; Helen Hartley as superintendent of nurses; and Harley Pettit as general superintendent and business director.

Plans for a new general hospital were announced by Pettit in December 1914. The city already had Ryburn Memorial Hospital, which opened in 1895 at Clinton and Madison streets. This new hospital was to be built on the grounds of the Ottawa Tent Colony on the south side (perhaps it was needed because many Ottawans have always regarded the south side of Ottawa, across the Illinois River bridge, as an entirely different place).

This was to be a general hospital. There would be a large administration building with an operating room, consultation rooms, a laboratory, offices, waiting rooms, and sleeping quarters for nurses. The two wings would have a capacity of fifty patients, and cots for more. Open-air treatment was part of the therapy. Each room would be connected to a veranda so the cots could be pushed out for fresh air. Ottawa architect Jason Richardson was given the contract.

It was estimated to cost between $50,000 and $75,000 and would have all the latest appliances and features. "The designs show an elaborateness possessed by few hospitals in this section of the state," the newspaper reported. "The buildings will be Mission in style and will cover a vast quantity of ground. They will embody only the latest ideas in hospital construction."

The story in the *Daily Republican* said that it would not conflict with Ryburn Memorial Hospital. There would be no emergency cases, and it would specialize in sanitarium care for a number of ailments.

It was named Illinois Valley Hospital, and it opened in 1915 as a diagnostic hospital with the latest in laboratory facilities. The intention was to keep the patient in their own community with their own doctor available for consultation rather than sending someone to a faraway clinic. The hospital was built to be more like a home than a hospital.

An early brochure showed rates of $30 to $45 per week for hospital services. Semi-private rooms were $4 a day; $6 a day for a private room; and $6.50 a day or a room with a bath. The major operating room fee was $7.50; the minor operating room was $5. For complete obstetric care for ten days, the fee was $50, or $6 a day if the stay was less than ten days. Special nurses cost another $2.25 a day. Meals were served at the Club House. Guests staying with a patient were charged $1 a night for a cot, plus fifty cents for breakfast and seventy-five cents each for lunch and supper. Dinner on Sundays and holidays cost $1.

Harley Pettit was president of the hospital and Dr. Royal Dunham was medical superintendent, the same positions they held at the tent colony. The hospital was renamed Ottawa General Hospital in 1933. Ottawa Arthritis Sanatorium and Diagnostic Clinic was added that same year by Dr. E. C. Andrews, who had graduated from the Kirksville College of Osteopathy and Surgery in 1919 and moved to Ottawa to set up a medical practice.

In the 1920s and 1930s, Illinois Valley Hospital was known as the "company hospital" for the Radium Dial Company in Ottawa. Women who worked at Radium Dial, painting clock dials with radium to glow in the dark, developed radium poisoning. Radium goes to the bone. These women had their teeth fall out, their jaws disintegrate, and they died horrible, painful deaths. Radium Dial had Ottawa doctors lie on death certificates, certifying that the cause of death was diphtheria, pneumonia, syphilis, and every ailment other than radium poisoning. It was a shameful part of Ottawa's past and on the name of the Ottawa medical community, and it marred the otherwise sterling name of Pettit.

A story in the *Bloomington Pantagraph* in June 1916 noted that Ernest Reutter, a Fairbury barber, took Roy Farney to Ottawa to become a patient at the tent colony. In October, Reutter was murdered by Eli Limeberry, a former employee. Limeberry shot Reutter to death at the corner of Maple and Fourth streets in Fairbury because Reutter had fired him for being drunk.

Chicago's health commissioner, Dr. John Dill Robertson, made a public plea in September 1917 for a "philanthropist to open a subdivision fitted with frame cottages at cheap rents for the care of tubercular patients."

Robertson said it could be self-sustaining because it would bring 5 or 6 percent on the investment. Cottages could be rented for $8 or $10 a month. Robertson also said his office made a city-wide survey and found there were 14,000 cases of tuberculosis in Chicago that had not been reported.

Dr. Pettit was named to a three-man council in November 1917 in charge of the 1917 Red Cross Christmas seals drive. More than 43 million Christmas seals were used in Illinois that year. Christmas seals was started in 1907 by the American Lung Association to raise money.

Officials in fifty counties across Illinois had voted to establish T.B. clinics in their area by 1921. A train carrying 100 people from these counties left Springfield on May 8 on a tour of sanitariums across the state. The caravan, sponsored by the Illinois Tuberculosis Association, was designed to give these officials first-hand knowledge on how to build and operate sanitariums. The group visited facilities in Springfield, Riverton, Bloomington, Ottawa, Peoria, Mackinaw, Quincy, and Jacksonville.

Fred Gunderson, a sixty-year-old employee at the tent colony, was a Swedish immigrant who had been in America for thirty-nine years, working a number of jobs before taking a job as a dishwasher at the tent colony in 1921. He preferred to bunk in one of the wooden huts rather than live inside the main building, which he was offered. He came to work drunk on the night of January 1, 1923, so he was sent back to his hut. He fell asleep while smoking in his bed. A fire started and the small hut was consumed before anything could be done. His charred body was found in a hunched position in a corner.

The 1930 U.S. Census shows Harley Pettit was the manager. Thor Rivers was the physician. Nurses were Anne Morris, Velma Bounds, Beryl Carter, Alice Stogdill, Grace Wellman, Mable Cousens, Hazel Watts, and Carrie Kuhn. Ora Woolley was a waitress. Katherine Einspater was a maid. Laborers were John Gower, John Hapeman, John Swanson, Thomas Burke, James Hogan, and Arthur Johnson. Only five people were listed as patients.

Joseph Weygand was a thirty-five-year-old waiter at the Ottawa tuberculosis sanitarium's dining hall. He was despondent because his girlfriend would not marry him, so he drowned himself in the Illinois River on July 26, 1936.

The pictures on the following pages are from a booklet on Illinois Valley Hospital, which became Ottawa General Hospital.

Ottawa General Hospital and the Ottawa Arthritis Clinic.

Ottawa Arthritis Sanatorium and Diagnostic Clinic, 900 East Center Street, Ottawa, Illinois

The west annex.

The east annex.

ROSWELL T. PETTIT, M. D.
OTTAWA, ILLINOIS
ILLINOIS VALLEY HOSPITAL

C O P Y

Chicago, Illinois,

August 27, 1929.

Dr. R. T. Pettit,
Ottawa, Illinois.

Autospy, Margaret Looney, Ottawa, Illinois.

"Autospy: Anatomical findings:
          Extensive diphtheritic pharyngitis
          and laryngitis.
          Marked edema of the glottis and
          epiglottis.
          Disseminated terminal broncho-
          pneumonia of both lower lobes of
          lungs.
          Terminal edema of the lungs.
          Dilitation of the heart.

"Microscopic Findings: Diphtheritic exudate
          with deep necrosis of the pharynx
          and larynx.

          Early state of broncho-pneumonia
          with diphtheria bacilli visible
          in the lung tissue.

          Degenerative changes in the heart
          muscle.

          Passive congestion of the liver and
          spleen.

"Smear from the pharynx made on August 10, 1929
is positive for diphtheria."

          Very sincerely yours,

     SIGNED: AARON ARKIN, M. D.

This is the first page of a four-page autopsy report on Margaret Looney, an Ottawa woman who worked at Radium Dial and who died of radium poisoning in 1929 at the age of twenty-four. The top page is on the stationary of Dr. Roswell Pettit, Illinois Valley Hospital, and shows the cause of death was diphtheria and other ailments—but not a mention of radium. Ottawa doctors were paid by Radium Dial to lie on death certificates to deny radium poisoning. The autopsy was done in the Dwyer funeral parlor in Ottawa by Dr. Aaron Arkin of the Chicago Laboratory Co., where Arkin was an officer. Not coincidentally, Radium Dial had an office was in the same location as Chicago Laboratory, in the Marshall Field building at 26 E. Washington Street in Chicago. Fifty years later, Miss Looney's remains were exhumed and examined at Argonne National Laboratory and were found to be "honeycombed" with radium, at a number that was more than 1,000 times above the level considered safe. There was so much radium in her bones that strong readings from a Geiger counter were detectable at ground level before the body was exhumed.

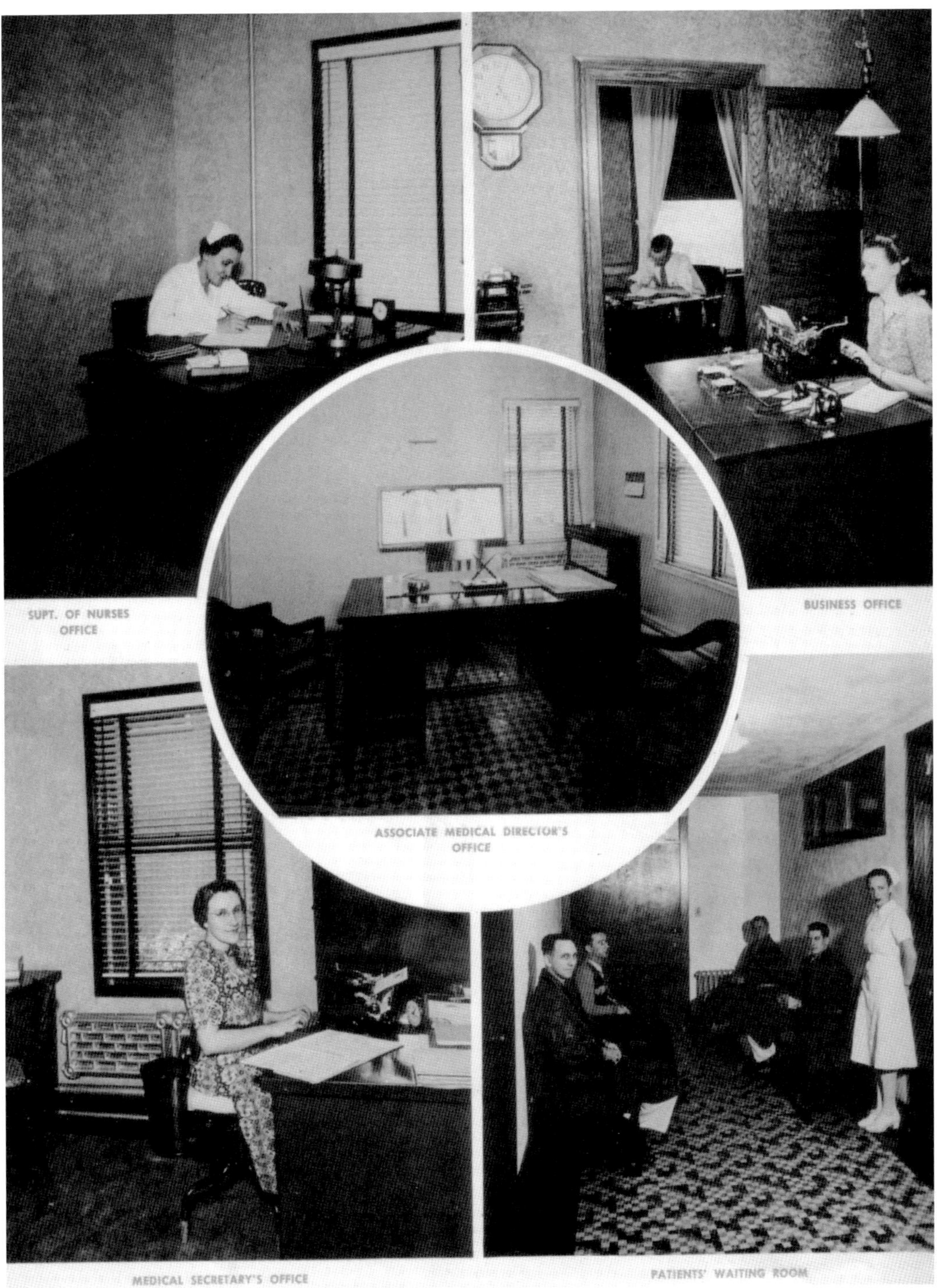

The pictures on the following pages are from a booklet on Illinois Valley Hospital, which became Ottawa General Hospital.

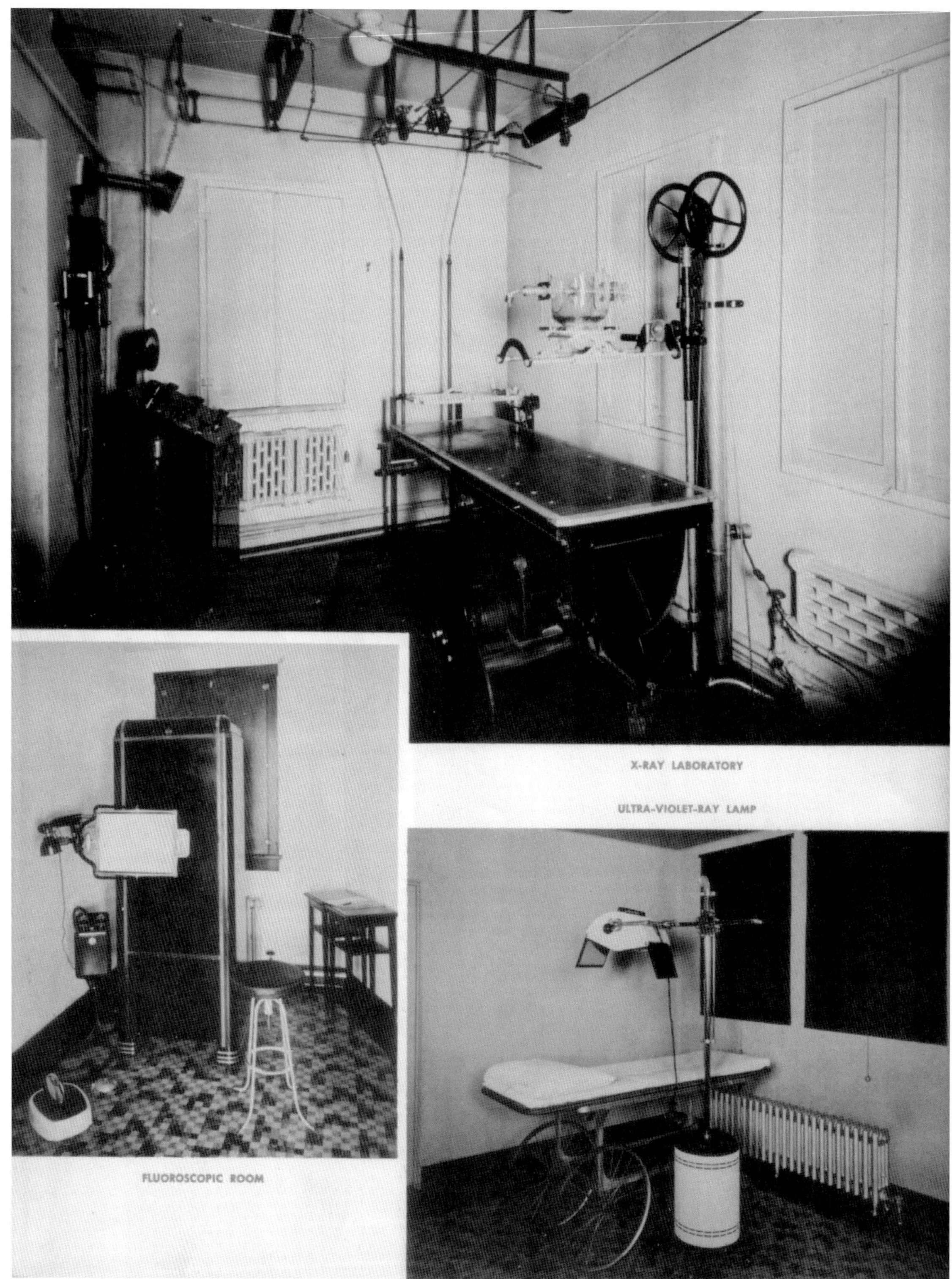
X-RAY LABORATORY
ULTRA-VIOLET-RAY LAMP
FLUOROSCOPIC ROOM

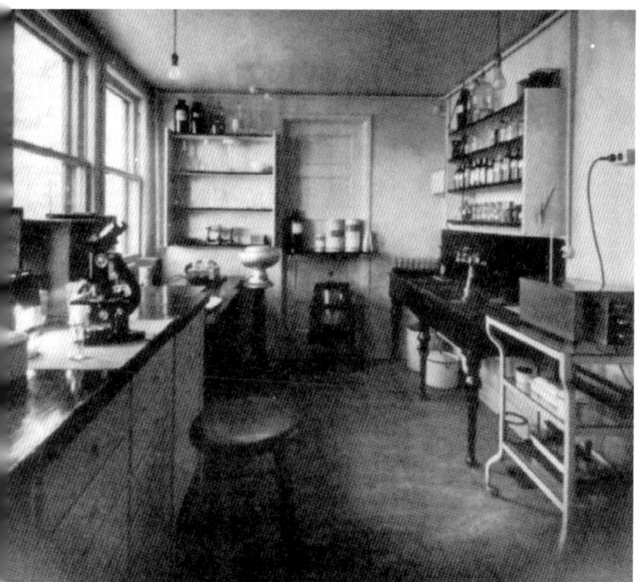
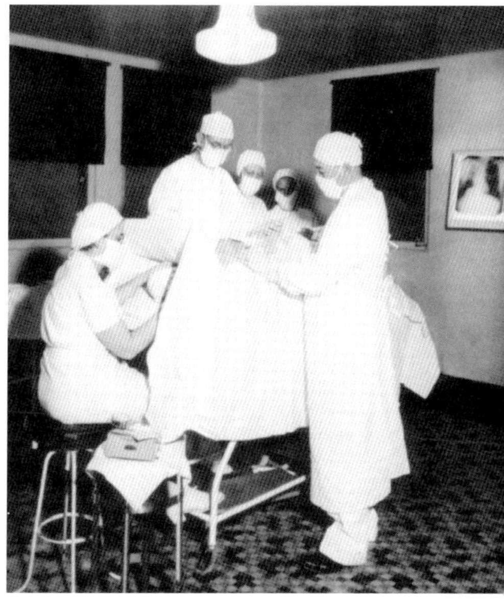

Above on the left is the clinical laboratory in the LaSalle County Tuberculosis Sanitarium. Above right is an operation in progress. These pictures are from a 1940s booklet printed by the sanitarium.

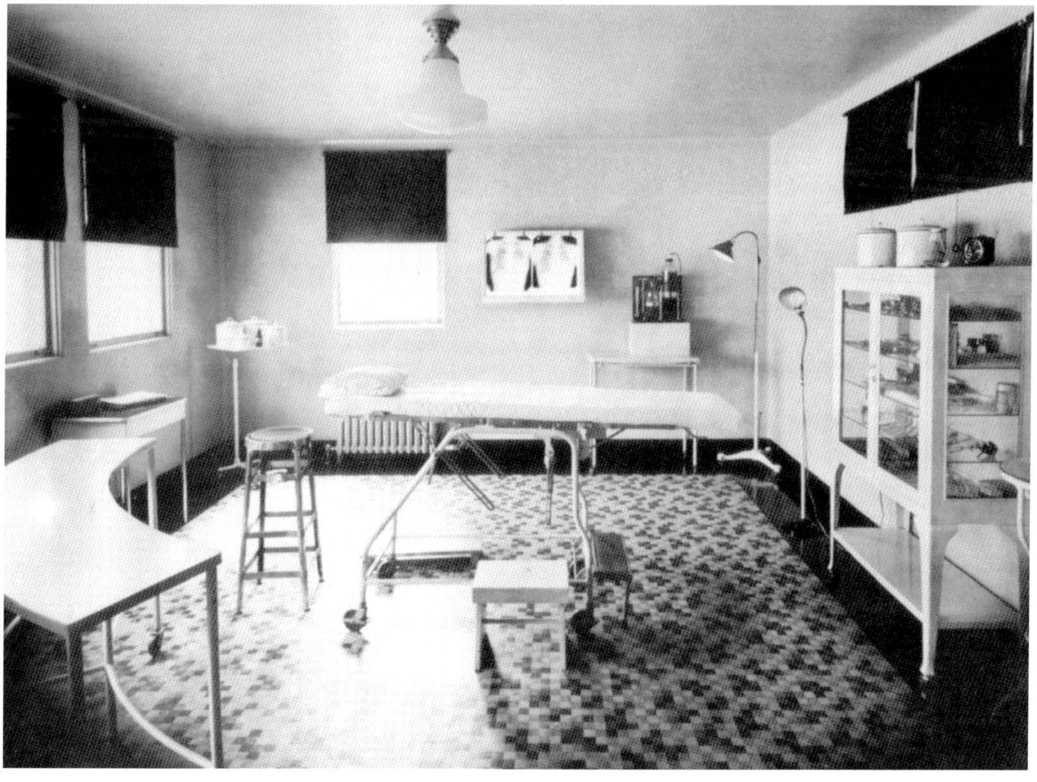

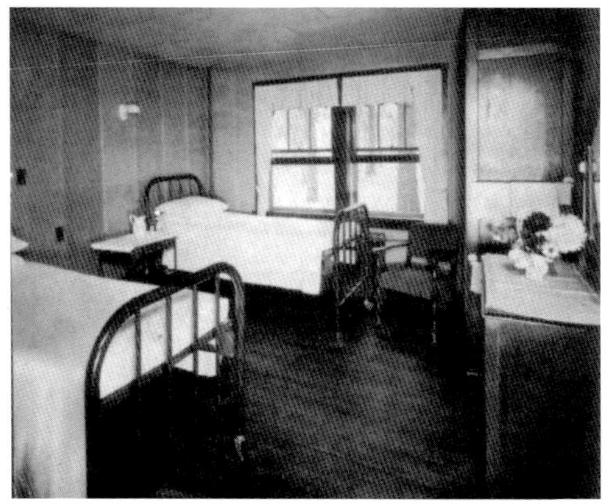

A semi-private room in the east annex.

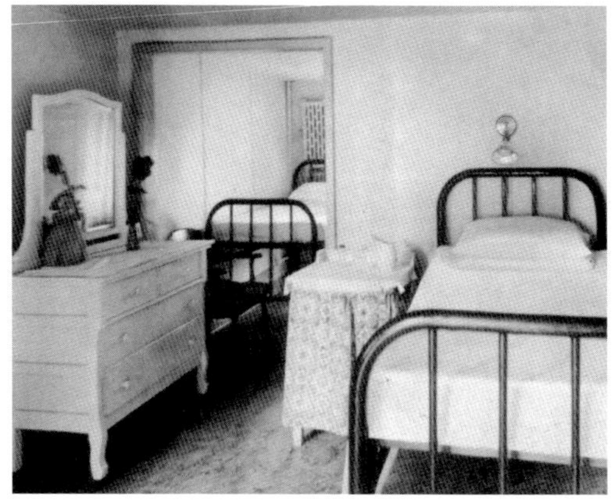

A semi-private room in the main building.

Surgical utility room.

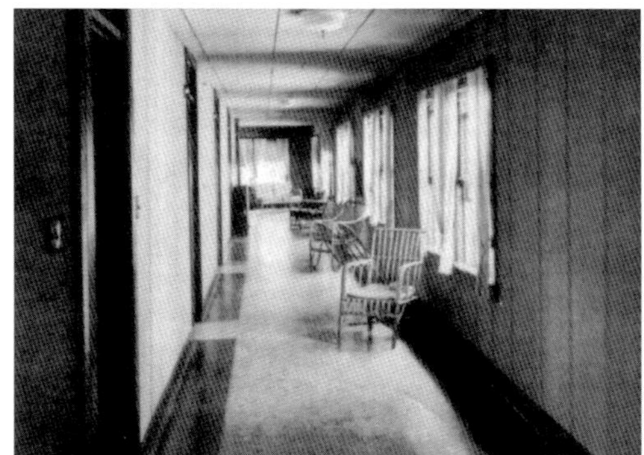

East annex corridor.

The patients' sun porch in the west annex.

3

Life in the Tent Colony

There is nothing better than a first-hand account, and there are a number of contemporary accounts of what life was like in the Ottawa Tent Colony and how the colony worked.

Dr. Pettit published a thirty-two-page book in 1905, which is descriptive of the colony with a number of important pictures that show what life was like in the tent colony. The pages from that booklet are at the end of this chapter.

There also was a pamphlet that gave the schedule for patients. A wake-up bell rang at 7 a.m., followed by a bell for breakfast at 7.30 a.m. There was a rest period after breakfast, followed by exercise from 9.30 to 11 a.m. There was lunch at 10 a.m. and dinner at noon. Another rest period was from 1 to 3 p.m. There was another lunch, and temperatures were taken at 3 p.m. Another exercise period was from 3 to 5 p.m. Supper was at 6 p.m. All lights were to be out and windows were to be opened at 9.30 p.m.

"Never stay out of doors while you are chilly, but learn never to be chilly out of doors," the pamphlet said. The head nurse was to be notified immediately if a patient had difficulty keeping warm. Patients were expected to live outdoors twenty hours a day and needed to dress appropriately. Woolen under-clothing with long sleeves was advised, but chest protectors were not allowed. Men should wear woolen shirts and corduroy trousers. Women's dresses should be loose-fitting with high collars and long sleeves, with a woolen petticoat or bloomers. Hoods or caps should cover the ears and forehead. Mittens were preferred instead of gloves. Preferred footwear included woolen socks, low-heel sheepskin shoes, and leggings that buttoned up to the knee. Rubbing your feet in the morning and at night was good for keeping them warm.

The *Ottawa Free Trader* told how the patients were surviving below 0-degree weather in a December 30, 1904 story:

While the blizzard was upon us the other day, you were probably very busy with fire and furnace. So busy that you probably never gave a thought to the inhabitants of the Ottawa Tent Colony on the bluff south of the city. If you did think of them, you probably threw in an extra shiver or two and permitted great gobs of sympathy to well up in your heart.

All of which was uncalled for and unnecessary. While the wind was blowing a gale of 70 miles an hour, the patients over there stuck to their tents. And they suffered not at all. In fact, they probably suffered less than the pampered householder who had to sneak downstairs two or three times a night, with the boreal zephyrs playing a wild, weird symphony through the abbreviated narrative of is *robe de nuit* to poke up the sitting room fire.

And here comes in the practical application of this experience. At some time in the future, sooner perhaps than you expect, the state will establish a sanatoria for the treatment of tuberculosis. The state has numerous other state institutions for the treatment of other classes. And on an average it costs $350 per person to house these wards of the state. To house a tubercular patient in a tent in this climate will cost but $35. In other words, for what it would cost the state to provide housing for one patient at such an institution as Kankakee, it could provide accommodations for ten at the sanatoria for the cure of tuberculosis.

Everywhere the experiments now being carried on at Ottawa are attracting widespread attention. And they are being closely watched by the medical fraternity. The National Society for the Study and Prevention of Tuberculosis is to meet at Washington in May of the coming year. Dr. J.W. Pettit, supervisor and director of the Ottawa Tent Colony, has been invited to address that body of experts.

"The winter of 1904-05 was a specially rigorous one in this locality," Dr. Pettit wrote. "During that first winter, provision was made to house the patients during the cold weather. They all, however, of their own accord, remained in their tents. This during a winter where the thermometer registered on several occasions twenty-five and thirty below zero. Instead of suffering from the cold, they were comfortable and rather enjoyed the experience. Several of those who were accustomed to living in frame houses declared they would have been less comfortable had they remained at home. Even the new arrivals during the extremely cold weather insisted upon going into tents. This was believed to be too severe a test, but in no cases did the management have cause to regret yielding to the entreaties of the patients. Their action is the more remarkable when account is taken if the fact that many, if not most of them, had come from homes where it was difficult to drive them away from the vitiated and super-heated atmosphere of their badly ventilated houses."

The results, he added, were even better in the winter than in the summer.

"It was Deadly Cold," according to the headline in the February 17, 1905, *Ottawa Free Trader*. It was -30 degrees the previous night. William Morrow, a watchman at a Peru foundry, was found frozen to death.

The *Ottawa Free Trader* reported it was -9 degrees on February 8, 1914, with bitter winds. The temperature was -3 degrees on December 17, 1914. A story in the *Ottawa Free Trader* told how the twenty-nine hoboes in the city jail were treated to wieners, hot rolls, and coffee, but no mention was made about how the people in the canvas tents were facing the weather.

The conditions were not as primitive as one might imagine. Dr. Charles Fox Gardiner of Colorado Springs, the man credited with coming up with the concept of the tent/hut for T.B. patients in America, described his tents in an article in 1902. His tents were not of the same design as the tents in Ottawa. His design was based on the tepees of the Ute Indians. The tepees had a conical shape with ventilation at the top and space around the bottom to let in fresh air.

"I noticed that in their tepees of skin, the ventilation was nearly ideal." Gardiner's tents were warm inside while letting in fresh air, no matter the temperature:

My tent is of dark khaki twelve-ounce duck, stretched over a six-sided framework of wood without any centre pole and without pegs or guy ropes so that it stands firm, like a house. The floor is raised eight inches from the ground and is in six sections so that it can be easily moved. The lower edge of the wall is fastened several inches below the floor and one inch out from it all around; this is to insure at all times an inflow of air that is gradual and without draughts, since this inch space in a circular tent represents an area of 520 square inches, and the hole in the top for overflowing air has an area of some 177 square inches. In this way, the tent cannot be closed and is ventilated automatically and constantly.

The open space between the floor and the walls of the tent allows air to flow in at all times, while the hole at the top allows air to flow out all the time. In this way, the tent always ventilates itself day or night, whether the door is shut or not, whether the interior is heated or not, or in any weather. As the air has to turn a corner to enter the tent, it cannot come as a draught, and as it passes in through all the inch space encircling the floor, it enters slowly and without force, being evenly distributed but coming through, collectively, a large area. There are also small shutters so constructed that they can partially close the opening from within the tent in case of very high winds. The opening at the top of the tent is covered by a zinc cone which can be controlled by pulleys and rope within the tent, in stormy weather being drawn to within an inch of the tent roof.

Colorado Springs Tent & Awning Co.

SOLE MANUFACTURERS OF **Dr. Gardiner Sanitary Tent.** (PATENT PENDING.)

This tent is circular or octagon in shape, with a wall five feet, roof 2-3 pitch of diameter —namely, a tent 12 feet in diameter, has eight-foot pitch, five foot-wall, and an extreme height of 13 feet. The top terminates in a ring, 15 inches in diameter, to which the rafters are fastened. The floor is made in **four or** five sections and raised five inches from the ground.

The edge of same all around has a n air space of two inches covered with wire netting, thus securing a gradual in flow, without draft, of pure air.

The 15-inch opening on top forms an exit for heated and impure air. This opening can be covered by a metal cap, which can be raised or lowered by means of ropes.

Dr. Gardiner says the advantages of tent life are plenty of fresh air and diffused light, all of the 24 hours, taken without effort.

Above tent can easily be put up or taken apart and packed for shipment.

All leading physicians of Colorado highly commend this tent for invalids. You can see many of them in use at the Nordrach Ranch and Glockner Sanitarium, also on many lawns throughout the country.

We are also large manufacturers of Tents, Awnings, Wagon Covers, Horse Blankets and everything made of canvass.

Campers and Hunters Outfits complete, from tents down to cooking utensils. Dealers in Cotton Duck, Wrapping Twine, Rope, Camp Blankets, Hammocks, Camp and Lawn Furniture of every description.

WE MAKE A SPECIALTY OF BUILDING HOUSE TENTS.

Colorado Springs Tent and Awning Company,

13 and 15 N. Cascade Ave, COLORADO SPRINGS.

An advertisement for Dr. Gardiner's tents for T.B. patients from 1902.

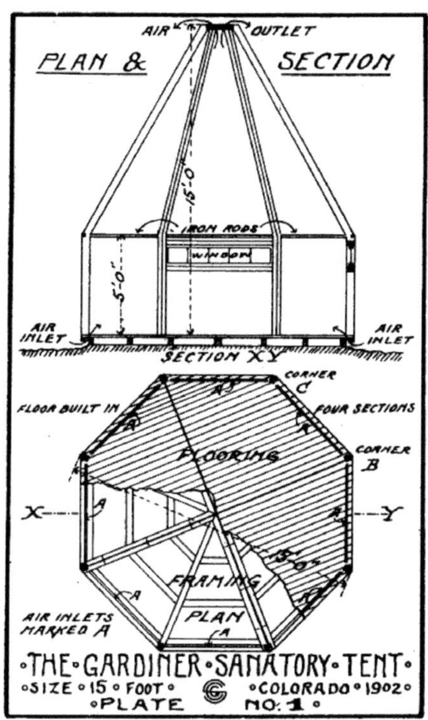
AIR OUTLET
PLAN & SECTION
15'-0"
IRON RODS
WINDOW
5'-0"
AIR INLET
AIR INLET
SECTION XY
FLOOR BUILT IN
FOUR SECTIONS
CORNER C
FLOORING
CORNER B
X
Y
A
FRAMING PLAN
15'-0"
AIR INLETS MARKED A
·THE·GARDINER·SANATORY·TENT·
·SIZE·15·FOOT·   ·COLORADO·1902·
·PLATE·   NO. 1 ·

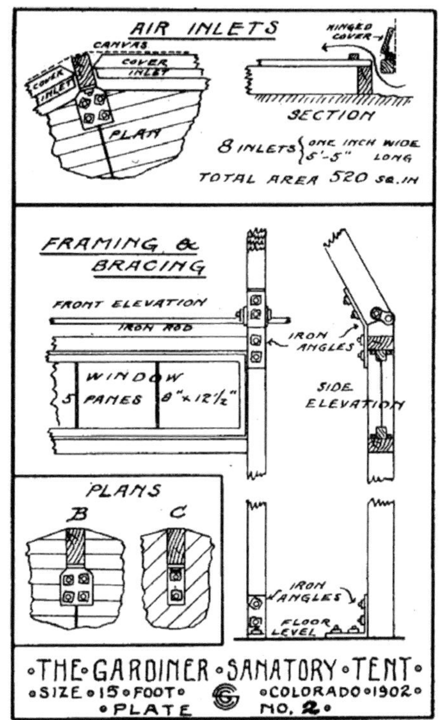
AIR INLETS
HINGED COVER
CANVAS COVER INLET
COVER INLET
PLAN
SECTION
8 INLETS { ONE INCH WIDE, 5'-5" LONG
TOTAL AREA 520 SQ. IN.
FRAMING & BRACING
FRONT ELEVATION
IRON ROD
IRON ANGLES
WINDOW
5 PANES 8"x12½"
SIDE ELEVATION
PLANS
B   C
IRON ANGLES
FLOOR LEVEL
·THE·GARDINER·SANATORY·TENT·
·SIZE·15·FOOT·   ·COLORADO·1902·
·PLATE·   NO. 2 ·

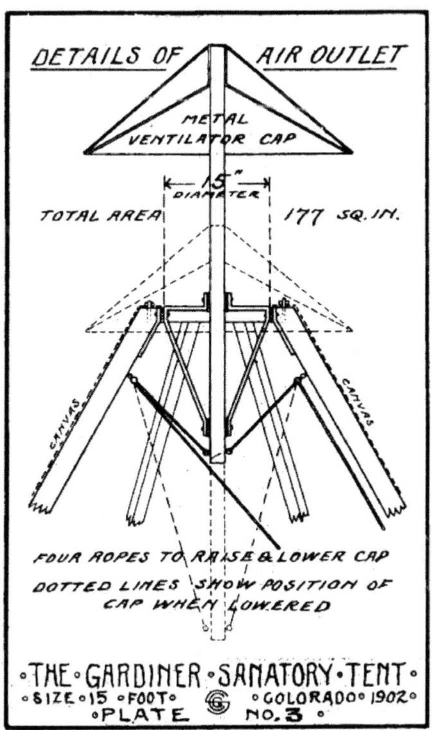
DETAILS OF AIR OUTLET
METAL VENTILATOR CAP
15" DIAMETER
TOTAL AREA   177 SQ. IN.
CANVAS
CANVAS
FOUR ROPES TO RAISE & LOWER CAP
DOTTED LINES SHOW POSITION OF CAP WHEN LOWERED
·THE·GARDINER·SANATORY·TENT·
·SIZE·15·FOOT·   ·COLORADO·1902·
·PLATE·   NO. 3 ·

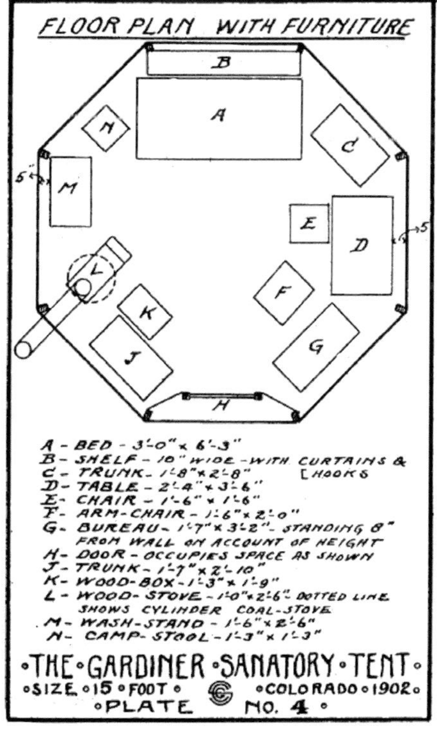
FLOOR PLAN WITH FURNITURE
B
A
N
C
5"
M
E
D
5"
V
K
F
J
G
H
A - BED - 3'-0" x 6'-3"
B - SHELF - 10" WIDE - WITH CURTAINS & HOOKS
C - TRUNK - 1'-8" x 2'-8"
D - TABLE - 2'-4" x 3'-6"
E - CHAIR - 1'-6" x 1'-6"
F - ARM-CHAIR - 1'-6" x 2'-0"
G - BUREAU - 1'-7" x 3'-2" STANDING 8" FROM WALL ON ACCOUNT OF HEIGHT
H - DOOR - OCCUPIES SPACE AS SHOWN
J - TRUNK - 1'-7" x 2'-10"
K - WOOD-BOX - 1'-3" x 1'-9"
L - WOOD-STOVE - 1'-0" x 2'-6" DOTTED LINE SHOWS CYLINDER COAL-STOVE
M - WASH-STAND - 1'-6" x 2'-6"
N - CAMP-STOOL - 1'-3" x 1'-3"
·THE·GARDINER·SANATORY·TENT·
·SIZE·15·FOOT·   ·COLORADO·1902·
·PLATE·   NO. 4 ·

Dr. Gardiner's "sanitory tent."

My tents are heated by central draught circular stoves which burn either wood or coal. And can be so regulated as to keep a good fire without care for ten hours. Even in zero weather, the tent can be kept perfectly comfortable to dress or undress in or to sit still and read. The stove is of such a size as to thoroughly warm the tent under any conditions, and at the same time it is impossible to over-heat the air or interfere with ventilation. The more heat used, the greater the displacement of heated air upward, and a more rapid interchange of air at once occurs. As the heated air can escape at the top, the fresh air can always enter at the bottom of the tent. This is automatic and not under the control of the invalid.

These facts I wish to emphasize, as they are of very great importance. The average invalid has an aversion to ventilation and will shut windows and in other ways embarrass the inflow of fresh air. In the sanitary tent it is impossible to do this as they are ventilated automatically.

The stove pipe passes out through the wall of the tent and by an elbow joint is carried upward to the height of the tent. It should be made of galvanized iron and can be guyed in place; it must not be below the top of the tent in order to insure good draught, and in burning wood, to lessen the danger from sparks.

A small window which does not open is used in these tents. It is placed horizontally and is one foot by six foot. It is most essential to have this tent furnished as completely and as comfortably as a bedroom. A proper bedstead, preferably of iron, with a thick mattress and plenty of covers; a shelf with hanging curtains on both sides and ends for hanging clothes; a wash stand and bureau; if possible, a commode; a box for coal and wood; a shelf for bottles and one for books; one or two good rugs on the floor; a steamer chair; an arm chair and a table for writing.

These need not be expensive but simple and strong. A scrap basket, broom, a dust pan, a hammer and some nails and a little saw frequently come in useful. In some cases,

This was the last streetcar trip to the tent colony on Ottawa's south side. Pictured are motorman James Beck, "Dad" Snyder, a patient, and Nellie Hulligan, a bookkeeper at the T.B. sanitarium.

I add a little zinc-lined box having two compartments, one of which is filled with ice and the other used to keep milk, eggs and etcetera cool. The lighting of the tent, if possible, should be by an electric wire taken from a nearby house; there might also be an electric bell to ring in the house.

In addition to the furniture belonging to the tent, I generally have made to use an outside awning, 12 by 10, stretched upon four uprights under which a hammock, chair and table can be placed. The tent is made of dark brown duck which does away with the white glare of light in the morning so disagreeable in most tents. The floor of the tent is about eight inches from the ground so there is very little fear of dampness.

Gardiner pointed out:

The inherent idea, common to many minds that during the cold months living in a tent must necessarily be a hardship is difficult to overcome. A tent generally brings up the picture in one's mind of cold, dust, wet, flies, glaring light and in general the usual experiences one has on a camping trip. But in a sanitory tent, especially if any degree of common sense and care is used to have the tent in the shade during the summer, these discomforts are not felt, and invalids assure me that they are quite as comfortable as in a house. I know many invalids living in the average boarding house room at health resorts who are certainly not as comfortable as they would be in the sanitory tent, either day or night. It is furthermore not a theory but a proved fact, this sanitory tent having been in actual use for two or three years and under different conditions of climate. Not only have incipient cases done well. But those far advanced with cavities have had fever diminished and general improvement occurs.

Dr. Pettit told a medical conference in 1905:

It sounds as if it were a hardship, sleeping in the open air, but patients never find it so when they actually undertake it. A man or woman living under such conditions acquires an enormous appetite, eating every day three regular meals, two lunches, from six to twelve raw eggs and drinking from five to six quarts of milk. This repairs the past and daily waste and builds up the patient.

OUT AT TENT PARK

This story was in the July 8, 1904 edition of the *Ottawa Free Trader*, describing the opening of the tent colony. It was headlined, "Out At Tent Park—Everything is Ready for Start of Experiment—Will be Watched With Interest." The account began:

Out at the point where Ellis and Highland Park butt into each other, and to the left hand of the streetcar track as you head toward Louis Hess' vaudeville stage, is one of the prettiest spots to be found on the banks of the Illinois River. A broad-topped, tree-covered bluff is skirted at one side by the river gently flowing while around the other end skirts a deep and picturesque ravine. It is an ideal spot for the purposes to which it has recently been dedicated.

In this beautiful grove has been erected the canvas cottages which the projectors believe will prove a haven of rest and restoration to those suffering from the preliminary stages of lung trouble. The recent developments of medical science seem to bear them out in their belief.

Tent Park presents a picturesque appearance. At one end of the street of tents is the big dining tent with table room for forty. Back of it is the kitchen tent. To one side of that, a well is being sunk that will provide an abundance of fine water. Besides which an ample supply of Ottawa's famous mineral water will also be on hand. At the other end of the street is the headquarters tent and the tent of the trained nurse.

The tents that will be occupied by the campers are of two sizes, all being provided with substantial board floors raised several inches above the ground. The 9 × 12 tents are for two occupants. The 7 × 9 tents are for a single camper. The tents are provided with a good iron bed, wire springs and mattress and a complete camper's toilet outfit. Campers furnish their own bedding. And even before the new institution was opened it was found necessary to order five additional tents.

The tents and grounds are lighted by electric lights and a row of incandescent bulbs will light the way out to the platform which is to be erected alongside the Ellis Park streetcar line. Swings, hammocks and settees will be scattered about the ground.

This is an energetic and scientific attempt to combat tubercular troubles by open-air living under the strictest sanitary supervision. It is a drugless institution in which no medicines are to be administered for the disease. Under certain conditions this malady is contagious, but under the strict discipline that will be maintained at Tent Park, it will be readily the safest point in all this surrounding country.

This will be the daily program of the campers: They will be expected to be ready for breakfast at 7 a.m., which will be the ordinary toothsome morning meal. At 10 a.m., each of them will be expected to surround a raw egg and a pint of milk. At noon, a substantial dinner. It will consist of the very best beef in some form, most of the vegetables, nuts and all of the fruits and ice cream. At 3 o'clock, another lunch of egg and milk. At 6 o'clock, a supper. Everybody must retire at 9 o'clock, after another lunch. There is a dietary program that would delight the heart of a newspaper man.

At noon today, the nine or ten campers who begin the treatment here will sit down to their first meal in the dining tent. They will be seated four at a table. These tables will have paper coverings with paper napkins, while as many of the side dishes as possible will be of the same material. These will be destroyed by fire. This is simply a sample of the rigid sanitary regulations that will govern the daily lives of the inhabitants of Tent Park.

They will be expected to spend every possible minute in the open air, but will not be permitted to exert themselves in any way. As for instance, down the winding road that leads to river, benches will be placed every 75 feet for the use of those coming up the hill from that direction. Croquet, lawn swings and other gentle bits of movement will "go." But lawn tennis, baseball, rowing, heated discussions of the primary law and all other violent forms of exercise will be "cut out."

The article continued, mentioning that the tent colony was part of the Illinois State Medical Society's crusade against tuberculosis. All the patients came from places outside of Ottawa, on the advice of their physicians. Only people in the early stages of the disease were accepted. Harley Pettit, son of the doctor, was in charge of the administration. The nurse in attendance was Marie Thompson and dietician Adelia Sater was in charge of the kitchen.

Postcards with messages on cards that are shown elsewhere in this book include one sent from Mrs. Christ Hendrickson to Mrs. George Johnson in Roland, Iowa, dated September 14, 1910:

Dear Cousin, We are all well. Have had such terrible rains now. A big barn burnt up of lightning 6 miles south of us Monday night. Was to the Kankakee Fair last week. Had a good time. The men are busy plowing. Haven't got our cows out of the muddy roads.

A card from Toots read: "This is a bit of the scenery up here. Of course, it is two [*sic*.] cold to take any boat rides but we'll send you a card just the same. The river is not frozen but the lake and the canal is."

On a postcard to Mrs. Fiedler in Winona, Minnesota, dated July 21, 1909, Mrs. Brown wrote:

Dear Mrs. Fiedler, I hope you read my letters to Mother B. and Mrs. Baker. I am feeling fine and know I am on the gain. This is a delightful place and I am enjoying it very much, only wish I could see my boys occasionally. I had a card from Mrs. Baker today from LaCrosse. Trust you are all well.

It is always a bonus to find postcards with personal messages. Toots wrote, "This [is] a bit of the scenery up here. Of course it is two [too] cold to take any boat rides but I'll send you a card just the same. The river is not frozen but the lake and the canal is."

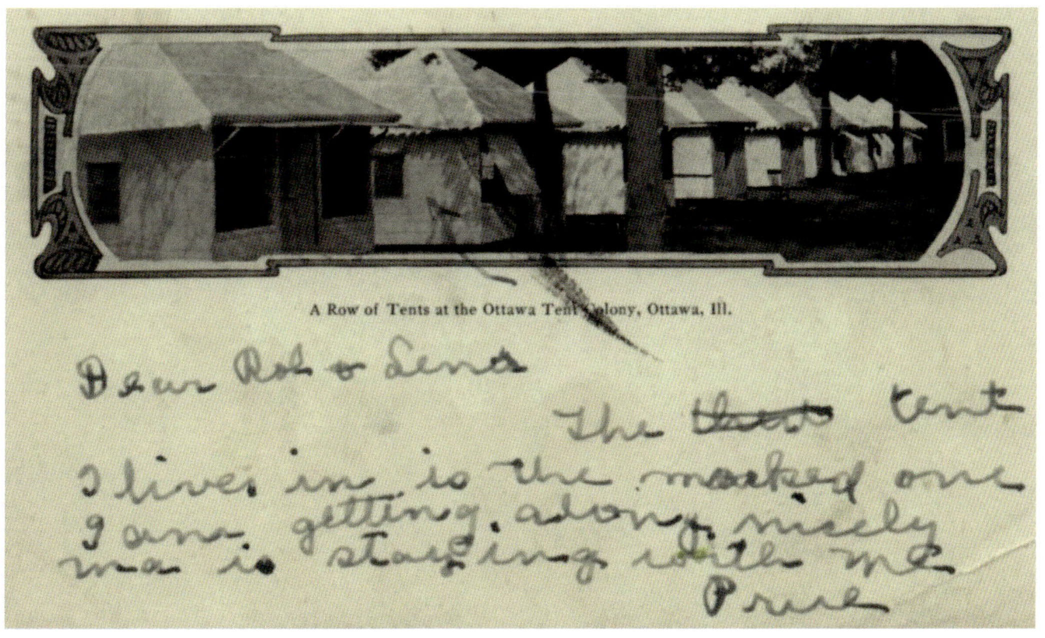

A Row of Tents at the Ottawa Tent Colony, Ottawa, Ill.

Prue wrote, "Dear Rob and Lena, The tent I live in is the marked one. I am getting along nicely. Wm. is staying with me."

Two panoramic views.

Postcard from Dorothy Berg showing the Club House with the dining hall: "It is a very nice place here and very nice people. Gained 3.5 lb. since I came. We stay outdoors all the time. This is the dining hall."

A FIRST-HAND ACCOUNT

R. Empey Frank, a patient in the tent colony, wrote a letter published in the *Chicago Daily News* in May 1912, describing life in the colony. It was reprinted in the *Bloomington Pantagraph* and the *Ottawa Free Trader.* The letter is lengthy, but it gives an invaluable look into the routine from an insider:

The northwest wind raced across a stretch of low country, over the frozen river, tore over the ridge of high land and swept howling away in the prairies to the south.

The canvas walls of the tents fluttered and tugged in their stays as the full force of the wind struck the little group of them snuggled together on the bluff overlooking the river.

At the Club House, the large windows rattled and shook in their sashes, and the snow swept madly across the long porch, the thick, black smoke from the kitchen chimney was driven before it had left the chimney mouth; but inside the house all was cozy and inside the little tents or in front, sheltered from the wind, the patients were snugly taking the cure.

The wind, the storm, the cold were a matter of course. The colony was prepared. A long dreary winter had settled down and the colonists were making the best of it. Day after day they would follow methodically the simple routine, and it was up to the weather to make things interesting.

This winter in particular it did, as I observed by day from the depths of my blankets as I lay outside in my steamer chair and by night in bed from beneath a heap of comforters, wool blankets, paper blankets, to which in some instances were added my overcoat, bathrobe and steamer rug, in my 9 by 12 tent which boasted the same temperature as that out of doors.

Meal time, this was an important occasion, for it was only for meals and for afternoon temperature that we enjoyed the comfort of a warm house, and for an hour or so we lived like other people. A stranger entering the dining hall at one of these important occasions could hardly believe that we were consumptives. Most of us looked so well, much better than the average person in the village or the people who caught a glimpse of the colony from the car, and who looked at us so sympathetically and yet doubtingly whenever we rode to town to do our shopping.

We were a happy lot, seemingly free from care and worry, for therein lay half the cure—not to worry. So we fostered cheeriness and good fellowship till it became, after a while, a happy spontaneous cheeriness instead of being premeditated.

What friendships have been woven together here, here where one needs encouragement and the help of friendship. I don't know whether it was because we were brought closer together thru sympathy and understood each other better because of the disease we had in common, or because each one took it upon himself to pass around as much cheer as he could, or what, but the spirit of the colony was remarkable, and the good naturedness practiced by both management and patient made the long, cold days pass swiftly.

Ah, those nights were good! Those crisp, clear, cold starry nights, just ideal nights to chase the bugs. Why, with every breath I could almost hear those poor TB's say, "What chance have we got in weather like this?" And the quiet, too, the stillness, how I enjoyed being out in the country, away from people—crowds of people, the rush and clamor of city life.

Why, the impertinent honk of the taxi, the noise of the cars, the clatter of the city's traffic that I had heard for years, never once awoke within me as much interest and enthusiasm as did the crunch of the frosty snow under some solitary foot steps, as I lay in bed and looked out into the night.

It was grand; it was just what I craved, to have a rest right next to nature. I thought of this, then. I had plenty of time to think, for often I found it too cold to read, my fingers almost froze through my heavy gloves as I sat blanketed in my chair.

The electric pad kept the feet warm by day, warmed the bed while the patient was at the house for supper, and worked faithfully all night from the head of the bed to the foot. I had a hot water bottle and a soapstone beat a mile. Often have I been thankful for it; when I wake up in the morning early, feeling the least bit chill, I only had to sneak out an arm, turn the switch and in a minute the juice would be sizzling thru the wires to the pad, to route the advance guard of a chill.

Cold feet—with an electric pad and fleece-lined shoes? I should say not. Cold feet are a detriment to the cure; warm must the feet be kept even to the sacrifice of appearances, which even the women made as they plodded along the narrow board paths to the Club House wearing those large sheepskin shoes over their street shoes. In their chairs, the "awful big feet" were hidden under the blankets.

In the dining hall at meal time, the patients sat three or four to a table, the women had their separate tables and the men their own. Here, their appetites having been whetted by the crisp, cold air, we did justice to our wholesome meals.

There was an art, or rather a knack, of getting wrapped up in our blankets, even with the help of a nurse, but to crawl in successfully alone was quite an accomplishment, for one had to first wrap up in a large horse blanket, place the electric pad at the feet, tuck the blanket tight to keep out the wind, then after the first blanket had been securely gathered around, to throw a smaller one, usually a steamer rug or a lap robe, over the legs and feet, recline in the steamer chair, at the same time gathering snugly around the shoulders.

In the Kiosk, or Ki house as we called it, a shed-like building open to the south and sheltered from the bitter north wind, the men usually sat. Here they took the cure together. Here farmers, clerks, machinists. Men from all walks of life spent hours and hours of inactivity but still at work in the business of getting well. Here good fellowship abounded, here each one did his best to make the hours seem the shorter, and to soften the rigid routine of the colony life.

It was a simple life. Rest, fresh air and wholesome food, the three fundamental principles of the cure were worked out to a science, and it was up to us to follow it out.

In bed with the tent flaps up by 9:30, up in the morning at 7:15, after breakfast to our chairs and blankets till the next meal—or if we were upon exercise, we took the allotted time, usually two by two, in walking leisurely and slowly at the "colony gait," thru the snow-blown fields or among the trees, along the paths down by the river; then after our walk back to our chairs. The afternoon the same—just a good quiet rest in the clear, crisp, crunchy, wintery out of doors.

It was no hardship. We were often given sympathy where none was needed. People spoke of us during that long spell of below-zero weather in such a sympathetic tone of voice and said, "Oh dear, I don't see how you stand it in such weather as this, how sorry I feel for you out in those cold tents." But we didn't mind it. We saw no reason why people should feel sorry for us. We played the game as determinedly and as conscientiously as we knew. We were happy, contented and comfortable. We were fighting for our health—and winning—we laughed at the winter.

(Signed) Ottawa, Ill. R. Empey Frank.

THE DOCTORS GIVE THEIR ACCOUNTS

Dr. Harold Moyer wrote a lengthy article in the *Chicago Daily News* in September 1904 about Ottawa's tent colony:

The tent colony of Ottawa has attracted widespread attention among the medical profession and the public of this and adjoining states. With one exception, it is the first tent colony for the treatment of tuberculosis established in the Mississippi Valley. More than 50 years ago, the value of the outdoor treatment was established in Europe, and some 30 years ago Dr. Trudeau established a sanitarium in the Adirondacks which has since grown famous. During all that time, little attention was devoted to tuberculosis. Patients, if treated at home, were given general advice and tonics, but the majority of those who had sufficient means were advised to try the effects of climate. Somehow the idea had become prevalent that there was a magic influence in the atmosphere of southern California, Colorado or Asheville. Patients were advised to avoid the severe winters of the northern states and go to the South. A few who went early enough while they had the strength to remain out of doors and digest poorly cooked meals, recovered. The majority of them either died away from home or had just sufficient strength to return home and die.

A full recognition of these facts led the Illinois State Medical Society, through its committee on tuberculosis, to establish the tent colony at Ottawa under the charge of Dr. J.W. Pettit. The purpose of this was to show that Illinois had just as good a climate for the treatment of the disease as any other portion of the country and that patients could be successfully managed near their homes. The movement met with instant recognition on the part of the profession. Though the colony has been established only a few months, it has already received a considerable number of patients, all of whom have made satisfactory improvement.

One patient, whom the physicians speak of as having been in the third stage of the disease, has made a marvelous recovery. She was very much reduced in strength and could scarcely walk. Within a few weeks the improvement was so marked that she was able to play croquet. While the success in this case is most gratifying, the tent colony has not been organized for the purpose of caring for very advanced cases, and it is not advised that they be sent. Those whose disease has not progressed so far as to make them bedridden and while they still have sufficient strength to be up and about are the ones for whom the colony is primarily intended.

While the principles upon which the colony has been organized have been accepted by the medical profession for a number of years, the public has had such an unfounded prejudice against the climate of Illinois that at first patients were loath to begin the treatment in this region. The results have proven eminently satisfactory. In fact, the patients now under treatment have made a greater improvement than those which have gone to a supposedly more favorable climate. A surprising feature of the outdoor treatment demonstrated in the current summer has been the rapid improvement in weight in nearly all cases. At first it might be supposed that the enforced rest would be irksome but this is not the case, as all the time is passed in the open air. The movements of the patients are alone restricted. The principles of the treatment are based on rest, feeding, open air and a maximum of sunlight. These are all equally important.

A want of rest and careful medical supervision has led to the failure of the climate cure, even when these are possible. The patient may be homesick or the diet may be such that the appetite fails. In such cases no benefit will be obtained from a change in climate. All these factors are met in the Ottawa Tent Colony. The diet is of the best, well prepared and daintily served. Taking it all in, the camp has a homelike and restful air. A visitor would scarcely believe that he was among a group of patients who were there for their health. Eggs, milk and cream, which of necessity form the staple articles of diet for the tuberculosis sufferer, are produced in abundance in the immediate neighborhood, the nearest neighbor to the tent colony being an exceptionally fine dairy.

No greater error can be made than to suppose that the tuberculosis patient needs a warm climate. The number of deaths from tuberculosis along our gulf coast is far greater than in the more northerly portion, in part this is probably due to a migration of consumptives to these supposedly favorable climates. Any climate that is enervating and moist is a poor place for a consumptive to go. Perhaps he would do better in such a climate if he were out of doors than if he were within doors in a colder climate, but the ideal atmosphere for a consumptive patient is cool or cold dry. Under such conditions, improvement is much more rapid, the appetite gains and the disease is often brought to an abrupt termination. It is not an infrequent experience in the outdoor treatment of patients in cold climates to have all the more severe manifestations of the disease disappear in two weeks.

The Ottawa Tent Colony will be transformed as cold weather approaches into a winter camp, in which the same principles of treatment will be carried out. It may not be practical to have a large number of tents, but suitable structures will be erected and the same plan of outdoor management will be carried out. Those who go there in the winter months will obtain a greater proportionate benefit from the same length of residence than those who have been there in the summer. This is the universal experience with outdoor treatment.

An assessment of the tent colony was printed in the *Wisconsin Medical Journal* in 1904, written by Dr. Arthur Patek. An excerpt follows:

The success thus far attained in the Ottawa Tent Colony has demonstrated beyond all doubt that tuberculosis can be cured in Illinois quite as easily and successfully as anywhere else. Nearly all patients admitted who were not in an advanced stage of the disease have been materially benefitted and several have already been discharged as convalescent.

The treatment consists in out-of-door life, a carefully selected diet, regulations of exercise and mediation for the improvement of nutrition. Each patient is expected to take no less than three quarts of milk and six raw eggs a day, in addition to the regular meals.

The results of tent life may be thus briefly summarized: The appetite increases; nutrition improves; cough decreases; night sweats cease; sleep improves; and the pulse rate is reduced. A marked improvement has been observed in all stages of the disease, but in early cases, the results are the most satisfactory. The patients are for the most part cheerful and contented with their surroundings. As a rule they accept uncomplainingly the primitive life which the treatment imposes. The greatest difficulty experienced is to hold some of them long enough to make improvement permanent.

A 1940s view of the medical building and the fireplace room in the Club House.

The Colony will be continued throughout the winter. Patients whose condition makes it desirable to live in tents will be recommended to do so. Those who should not, or will not, will be comfortably housed. A tent is intended simply as a sleeping apartment. Tent life as carried out for the treatment of tuberculosis soon becomes thoroughly enjoyable. Before beginning such a life, patients almost invariably entertain exaggerated ideas of its dangers and inconveniences. A short trial soon dispels this fear, and they are with difficulty induced to return to an indoor life. It should be understood, however, that during the winter, patients are given their choice, and those who object to living in a tent are comfortably housed in such a manner as will comply with the essential principles of the present day treatment.

The tuberculosis patient, when well fed and warmly clad, feels the exhilarating effects of cold weather quite as much as the normal individual, but will not avail himself of these valuable aids in the restoration of health unless under careful supervision. The Ottawa Tent Colony is a much needed and laudable undertaking and deserves the encouragement and support of the entire medical profession, without regard to school of practice.

A patient lying outside cottage number 65 at the Ottawa Tent Colony.

The 1905 booklet published by Dr. Pettit.

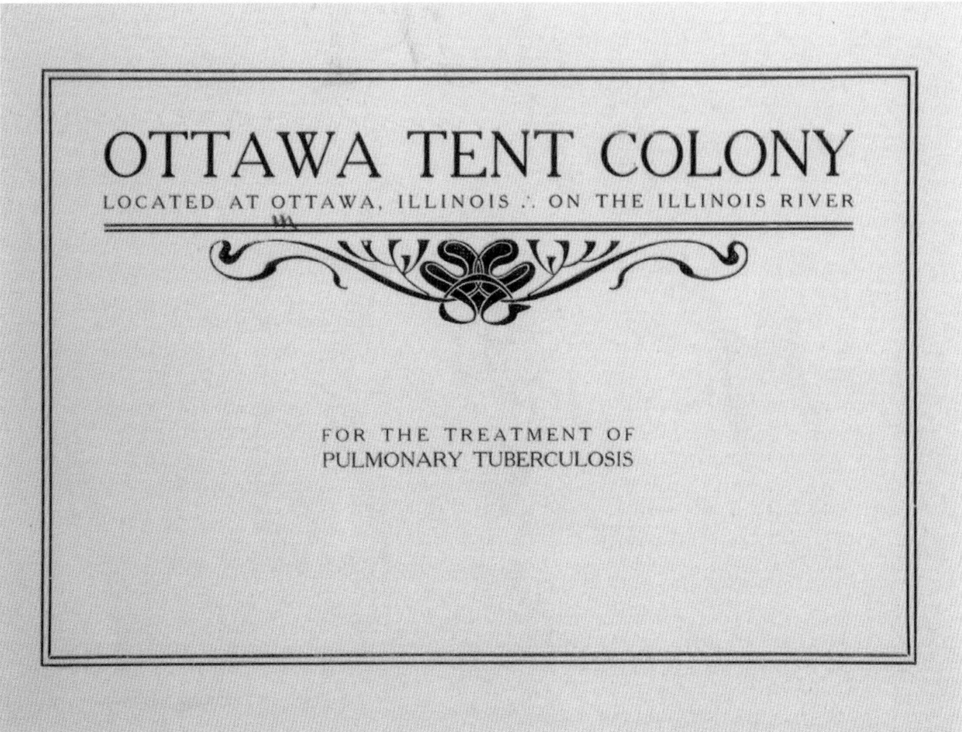

PERMANENT BUILDINGS AT THE COLONY

HE OTTAWA TENT COLONY is one tangible result of the campaign against the spread of pulmonary tuberculosis inaugurated by the Illinois State Medical Society at its May meeting in 1904. It was inaugurated as an experiment. Dr. J. W. Pettit, its founder, and still one of its directors, was pledged the moral support and encouragement of many of the representative physicians who constitute the membership of that society. The colony was opened in July, 1904, with two patients. It has grown since then. It has accomplished not all, perhaps, that mistaken enthusiasts may have expected, but it has done all that those who have intelligently studied the outdoor treatment of this dread disease hoped for at the beginning.

It has more than justified its existence, and demonstrated that it deserved to be put upon a permanent footing. It was not practicable to conduct it as a charitable enterprise, nor even, as in the early days, to furnish the treatment at actual cost. Into the equipment

CLUB HOUSE AND ADMINISTRATION BUILDING

STREET RAILWAY STATION

of the institution as it now stands has gone more than $40,000. It has been the aim to provide every convenience, every requisite, for the scientific application of the most modern methods of treatment. Yet care has been taken to avoid making the per capita cost of the plant so extravagant as to render the treatment prohibitive because of the expense. The physical equipment of the Ottawa Tent Colony is so well represented in the accompanying illustrations as to leave but little to take the form of printed words. There is a large

RESERVOIR

club house or administration building. Here live the resident staff of the institution; here is the modern, sanitary kitchen in which is prepared all the food; here is the lounging and rest room, with its artistic boulder-built fire place; here is the great dining room, open on two sides to every breeze that blows (save in the most inclement weather), with its mission furniture and its

DINING ROOM

FIRE PLACE IN LOUNGING ROOM

paper napery, destroyed after each meal. And sweeping around this building is the great, broad veranda, commanding a beautiful stretch of Illinois river valley scenery. Further along the river bluff is the modern bath house with every appointment perfect. Back through the woods and fronting a romantic ravine is a covered shelter, or kiosk, where patients spend many a pleasant hour in the open. Between these permanent structures, set in orderly rows, are the tents, from which the colony derives its name—all of the latest and most approved models.

All this is located on a beautiful wooded bluff, one hundred and twenty-five feet above "thy waters gently flowing, Illinois." Sweeping around one end of this great bluff is a deep ravine. Nature has done everything it could to make this an ideal spot for the purposes to which it is now devoted. Pure air, pure water and the most perfect drainage are here found in an ideal combination.

Club house, bath house, the tents, walks and drives are all lighted by electric light. No detail that can add to the comfort or convenience of the patients has been omitted. The work of beautifying the grounds is one that is constantly going on towards an ideal state.

DINING ROOM SIDEBOARD

BATH HOUSE

INTERIOR VIEW OF BATH HOUSE

The water supply of the colony is one feature to which special attention has been devoted in order to insure its purity as well as the adequacy of the supply. The tents in which the patients are housed represent the very best type of those structures as approved by the experience of this and other institutions of a like character. All these buildings and tents are located amidst pleasant surroundings. The grounds are kept in inviting condition and t h e entire atmosphere of the institution is one conducive to the restful spirit that makes for recovery in these cases. The bath house in connection with the colony is one that represents the latest thought in sanitary engineering. The same is true of all the equipment of the plant. Drainage is complete and perfect cleanliness is the marked characteristic of everything. This is especially true of the kitchens in which the food stuffs—a very important part of the treatment—are prepared. So thoroughly has this matter of perfect sanitation been gone into that visitors to the colony remark the comparative absence of house flies and other insect pests so noticeable in most public places.

AVENUE OF TENTS

DAILY LIFE OF THE PATIENTS

NLY conservative claims are made for the treatment given at this sanatorium. Its success largely depends upon a recognition of its limitations and requirements. The tent is intended chiefly as a sleeping apartment, and yet it is made an ideal living apartment, in which the purity of the air is maintained and the temperature regulated during all conditions of weather. Governed by well selected rules, tent life soon becomes thoroughly enjoyable, and the patient speedily loses the exaggerated ideas of its dangers and inconveniences. The patient is under constant medical supervision. He is removed from all disturbing influences. He is encouraged to take adequate amounts of food. He is made to take sufficient rest, and is prevented from over-exercise. He is taught how to live; he breathes pure air continuously, night and day;

GROUP OF PATIENTS

ON THE CLUB HOUSE VERANDA

ONE STYLE OF TENT

he is induced to take proper exercise; he lives with one object—that of getting well. The good effects upon others is a constant and encouraging object lesson. No one who is well enough to be up and about can fail to be benefited by the daily life at such a sanatorium as the Ottawa Tent Colony.

Sight must not be lost of the fact that these satisfactory results can only be obtained where plenty of time is given to the cure. The treatment must be carried out faithfully in incipient cases for at least three months and in advanced cases for at least six months, and often for a longer time, in order to obtain satisfactory results. The daily life at the colony does these things for the patient: Exposed continuously to fresh air, he gains an appetite, assimilates his food better, sleeps more soundly and awakens more refreshed. Life in the open air is the best agent for reducing fever. Night sweats usually cease very promptly and colds are practically unknown among patients living in the open air.

ANOTHER STYLE
OF TENT

In every way possible it is sought to make the daily life of the patients as pleasant and inviting as possible. To the other attractions of the dining room, that of music is frequently added. An orchestra of ability discourses music at regular intervals during the week while the meals are being served. Entertainment for the patients in other ways is provided, carefully avoiding all those that call for anything like tiresome effort. Frequently a pleasant evening is spent in the big dining room with musicians and vocalists of more than local repute. These evening entertainments include also illustrated travel lectures, as well as scientific lectures on the treatment of tuberculosis, with stereopticon slides to illustrate the various phases of the subject. While these addresses contain the latest thought on these subjects, special care is taken to make them interesting as well as instructive.

WINTER AT THE COLONY

N selecting the illustrations for this booklet which are to depict winter life in the colony care was taken to choose negatives which were taken during the winter of 1904-5—since which time the style of tents has been changed. These selections were made because the winter of 1904-5 was a specially rigorous one in this locality and the equipment meager; therefore the results obtained were all the more notable. During that first winter provision was made to house the patients during the cold weather. They all, however, of their own accord, remained in their tents. This during

OUR FIRST WINTER

"TAKING CURE"—FIRST WINTER

a winter when the thermometer registered on several occasions twenty-five and thirty degrees below zero. Instead of suffering from the cold, they were comfortable and rather enjoyed the experience. Several of those who were accustomed to living in frame houses declared they would have been less comfortable had they remained at home. Even new arrivals during the extremely cold weather insisted upon going into tents. This was believed to be too severe a test, but in no case did the management have cause to regret yielding to the entreaties of the patients. Their action is the more remarkable when account is taken of the fact that many, if not most of them, had come from homes where it was difficult to drive them away from the vitiated and super-heated atmosphere of their badly ventilated houses.

With the new tents now in use there is no thought of having the patients do aught else than sleep therein during the winter months. Before beginning such a life patients almost invariably entertain exaggerated ideas of its dangers and inconveniences. A short trial soon dispels this fear and they are with difficulty induced to return to an indoor life. Experience teaches

"TAKING CURE"—FIRST WINTER

that the results are even better in winter than in summer. The tuberculous patient, when well fed and warmly clad, feels the exhilarating effects of cold weather quite as much as the normal person, but will not avail himself of these valuable aids to the restoration of health, except when under careful supervision. In favorable climates the tent has been used very largely and successfully, but it did not occur to even its most sanguine supporters that it was practicable in unfavorable climates. At first thought it seems incredible that patients can be made comfortable in an ordinary tent at a temperature of twenty-five degrees below zero. Yet this is just what has been done at the Ottawa Tent Colony during one of the most severe winters experienced in the northwest for many years. It is irrational to house tuberculous patients in substantial buildings.

THE KIOSK

ON THE ILLINOIS RIVER

ITUATED as the Ottawa Tent Colony is, on the high bank of the Illinois river, a fine launch makes for the convalescent patients a playground, as it were, of the most beautiful portions of the Illinois river. And nowhere is there a more varied and entrancing stretch of scenery than for the ten miles along this stream between Ottawa and the historic Starved Rock. For the convenience and pleasure of patients there is provided during the boating season a launch well adapted for the purposes to which it is put. The "Fleur de Lis" is a part of the Tent Colony equipment which is an important adjunct to the treatment.

It is not the purpose of this little booklet to go into detail as to the scenic beauty of the Illinois; but simply to make mention in passing of some of the beauty spots that are of easy

LOOKING NORTHWEST FROM CLUB HOUSE VERANDA

access from the colony by means of the launch. In the list are included some scenes away from the river, but not to a distance so great that they may not be viewed by the patients without too much exertion or fatigue.

Patients of the Tent Colony need not, however, even board the launch to revel in scenic beauty. From the broad veranda of the club house itself, stretching away to the north and east, lies a land- and waterscape that arouses enthusiastic words of praise from every beholder. This particular point

RIVER WALK

of view furnished to Edgar Cameron, the well known Chicago artist, inspiration for one of his effective pictures in which the rich autumnal foliage of the Illinois valley is depicted in its true art values.

That view from the veranda is one that impresses every beholder. It is a reminder that before journeying down the river in search of scenic beauty one should see all in this line that the grounds themselves afford. The view down the ravine from the bridge is in itself a most romantic outlook. The river

LOOKING NORTHEAST FROM CLUB HOUSE VERANDA

FALL CREEK
BRIDGE

walks beside the slow, picturesque old stream open up vistas of enchantment that linger long in grateful memories. To sit on that broad veranda, with the moonlight streaming over the shadowed valley, with strains of music from an anchored launch at the foot of the wooded bluff rising to soothe the ear is a scene that never fails to charm.

It is at the foot of this great bluff on which the club house stands that one steps on board the "Fleur de Lis" for a trip down stream. One sails on down

the stream with the city of Ottawa on the right, passing the mineral spring grounds and under the bridges that span the river. To the south along that same river are winding roads that at every turn reveal new beauties. The low lying banks on the north side of the stream are in striking contrast with the higher and wooded bluffs that run close to the water line in many places on the other shore.

One glides on down the stream, by islands round which the placid waters ripple. Some of these islands are as nature left them, some are useful as they are beautiful, the

LOOKING EAST FROM TOP OF BUFFALO ROCK

78

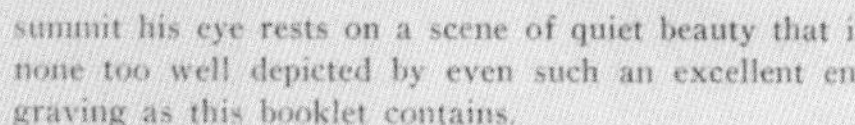

BUFFALO ROCK FROM THE RIVER

verdure which decorates them being the growing crops of waving corn.

One of the most striking formations along the river is Buffalo Rock, located on the north side of the stream and paralleling it with its lofty, rocky buttress for almost three-quarters of a mile. Striking as is this rock, when one leaves the launch and ascends to its summit his eye rests on a scene of quiet beauty that is none too well depicted by even such an excellent engraving as this booklet contains.

Further down the stream one rounds "Hungry Point" to land at the Horse Shoe Canyon. These narrow ravines are characteristic of the scenery along the Illinois. They are many in number and each has a beauty of its own. High up their sides grow the wild ferns that are of beauty and delicacy such as are no-

IN HORSE SHOE CANYON

where else found in all the state. Through most of these canyons run little streams that here and there break into little cascades and miniature waterfalls.

Further on down the stream one reaches a region where historic legend adds its charm to the wealth of nature's beauties. As one journeys down the stream towards the historic Starved Rock his eye first takes in the heights of Lover's Leap. The legend attached to this rocky cliff may have its counterpart in the folk-lore traditions of other places, but the place has a beauty and attraction that is all its own. Beyond this lies the jeweled centerpiece of all this riot of attractive surprises for the eye—Starved Rock—the Fort St. Louis of the early French explorers of the Mississippi valley.

IN SALT WELL CANYON

Down this valley went the great La Salle in 1670. He marked and noted this great rock, and a few years later Fort St. Louis crowned its heights. Three years after La Salle's first visit Marquette and Joliet were there. In those early days the

LOVER'S LEAP

great village of the Illini Indians clustered at the foot of Starved Rock. The story is told in many forms of how this place was the last stand taken by the Illini and that here the tribe was exterminated. These legends are by some pronounced a fiction. It is not true that at this point the last remnant of the tribes of the Illini was wiped out, for there are still living many of the descendants of that once powerful people. It is true that a war party of the Illini took refuge on the rock and were surrounded on all sides by the enemy. They were prevented from securing food or water and starvation did its fatal work. Only a remnant of the little band was left. The few survivors made at last a desperate dash in the middle of a dark and stormy night and some of them succeeded in eluding the foe. Most of them, being weakened by their long starvation, were unable to fight and were soon dispatched by their sleek and well-fed pursuers.

Among the remnants of the Illini still living there is a well defined legend or tradition, which is fully believed by them to be true, that this event was the last experience of their

STARVED ROCK

people on the east side of the Illinois and that, following it, they abandoned their entire line of defense along the river and never again crossed it. They feared to place its waters between them and the place of their retreat in case of defeat. It was more than a century after La Salle's visit, namely in 1780, that these events occurred.

Enough has been shown in this little booklet to more than make good the assertion that the old river flowing by the bluff on which is located the Ottawa Tent Colony can furnish both pleasure and profit to the patients. A trip down the river with a landing made at some one of the beauty spots for lunch constitutes an ideal afternoon's outing. And at that only brief mention has been made of the most striking points along the stream.

OTTAWA TENT COLONY STAFF

AT the head of the Ottawa Tent Colony is its founder, Dr. J. W. Pettit. He was placed in charge when the colony was established as a demonstration of the possibilities of fighting "the great white plague" by the Illinois State Medical Society. During this experimental stage he gave to the establishment freely of both his time and money. Then came a time when it was seen that if the colony was to continue it must become a private enterprise. At this period in the development of the institution Dr. Butterfield became associated with Dr. Pettit in its management. They are both graduates of Rush Medical College, being members of the same class in that institution.

Drs. Pettit and Butterfield have steadily and consistently refused the assistance of laymen in the conduct of the colony. They have taken this position because they desire to conduct it along strictly scientific and ethical lines, and do not wish to have it assume the

E. H. BUTTERFIELD, M. D. MEDICAL DIRECTOR	THE STAFF	H. V. PETTIT SUPERINTENDENT
MRS. T. G. HAYNES HEAD NURSE	J. W. PETTIT, M. D. MEDICAL DIRECTOR	MISS ADELLA SATER DIETITIAN

character of a purely commercial enterprise. The professional spirit and professional ethics are kept constantly in mind in the conduct and management of the colony.

The treatment is in strict accordance with the best modern information as to correct treatment of pulmonary tuberculosis. The colony has demonstrated the scientific and economic value of the tent in a cold climate.

One of the greatest problems in combating the disease is that of the proper nutrition of the patients. Not only that they shall have the best and most nutritious foods, but that this daily food shall be properly balanced so as to contain exactly the elements needed by the individual patient. This important department of the colony's work is in charge of an expert dietitian—Miss Adella Sater. Miss Sater is a graduate of the domestic science department of Lewis Institute. She is the first of her profession to devote herself exclusively to the line of work conducted at the Ottawa Tent Colony.

The important department of scientific nursing and care of the patients at the colony is well cared for. The head of this department is Mrs. Tamar G. Haynes, a graduate nurse of much experience.

The business management of the colony is in charge of Mr. H. V. Pettit. To the multitudinous details of the conduct of the establishment he devotes all his time and contributes

THE GUALANO ORCHESTRA

much to the comfort of the colony's inhabitants. The colony is located on the lines of the local street car company, and at the entrance to the grounds a platform with adequate shelter is provided. In the way of amusements for the patients much is provided, including the diversions of cards, chess, checkers, dominoes, and such like games, while outside recreation is furnished in the form of croquet and other amusements not calling for great exertion. All the latest newspapers and periodicals are kept on file. Mail for the patients is received twice and dispatched twice each day from the colony. To the permanent equipment of the colony is now being added a well appointed infirmary.

4

THE PEOPLE

James Wiley Pettit was born on August 4, 1848, in Sparta, Randolph County, Illinois. His father, Jonathan, was born in 1818 in Illinois; he was a furniture and cabinet maker. Jonathan married Mary Pillers in Randolph County in 1845. James had two siblings, Margaret and John.

Jonathan Pettit died in 1851. Mary married John Glenn in 1852 in Randolph County, and they had four children: Charles, Mattie, William, and Samuel. Mary died in 1870; John died in 1882. John Glenn was a doctor, which may have influenced James to enter the medical profession.

James Wiley Pettit enlisted in Company "K," 142nd Illinois Infantry, in June 1864. He was fifteen years old. The 142nd was a regiment that had a 100-day assignment, guarding railroads in Memphis. James then enlisted in Company "F" of the 154th Illinois Infantry in February 1865, when it was organized for another 100-day assignment. He served until September. He was mustered out in October. His military record shows he was 5 feet, 4 inches tall when he enlisted.

James, who always preferred to go by the name J. W. Pettit, graduated from Louisville Medical College in 1873. He began his practice in Ottawa in 1877 with an office at LaSalle and Jefferson streets. He lived in a house at Webster and Fillmore streets. Dr. Pettit earned a second degree at Rush Medical College in Chicago in 1884.

Dr. Pettit contracted a case of tuberculosis and moved to nearby Sheridan in 1884. He was unable to practice medicine for five years. However, he did purchase a sand mine there. He regained his strength and moved back to Ottawa in 1892.

When he returned in 1892, he built a fine house at 1401 Ottawa Avenue. Dr. Pettit sold the house in October 1904 and moved to 1017 Paul Street. He lived there until 1925, when the hospital built him a house at 823 E. Center Street, a block from the hospital.

Dr. Pettit's office was at 720 LaSalle Street in 1900. A few years later, he moved his office to the Moloney building at 112 E. Madison Street.

Dr. Pettit spoke at the annual meeting of the Illinois State Dental Society, in the Illinois Supreme Court building in Ottawa, in May 1875. The Supreme Court was located in Ottawa until it was moved to Springfield in 1897. The Ottawa building is now the home of the Third District Appellate Court.

J. W. Pettit.

Roswell Pettit.

Harley Pettit.

In addition to his work as a physician and director of the tent colony, Dr. Pettit was Ottawa city health officer from 1877 to 1881, and he was LaSalle County coroner from 1878 to 1880. He was chief medical officer at the county asylum and was president of the LaSalle County Medical Society. He became active in the newly organized National Guard and was a second lieutenant in Company "C" until 1898. He was appointed local surgeon for the Rock Island Railroad in 1896, and he was a trustee of the Illinois State Charitable Eye and Ear Infirmary in Chicago. The Pettits belonged to the First Presbyterian Church in Ottawa in the 1870s and he was secretary of the church board at that time. The Pettits later became active in the Congregational Church.

Dr. Pettit went on a Celtic cruise in early 1902 and wrote several letters about his trip. The *Ottawa Free Trader* published them all in a pamphlet which was sold in local book stores. "Dr. Pettit claims nothing for these letters but their truthfulness," the newspaper said. "This is by far too modest a claim, for there is not a dull paragraph or line in the entire series, but much that is of rare value and great interest."

A rally was held in Sheridan on October 23, 1902, and Dr. Pettit gave a speech urging voters to approve a referendum to reform primary election laws. Dr. Pettit spoke to a large crowd for an hour in favor of choosing candidates by votes instead of by caucuses and party bosses.

Dr. Pettit was president of the Illinois State Medical Society in 1908 and 1909. He also was president of the American Sanatorium Association in 1924 and 1925.

J. W. Pettit married Mary Louise Wilgus on September 24, 1874. Mary was born in Beloit, Wisconsin. She was a founding member of the Illini Chapter of Daughters of the American Revolution in 1896 and was elected president of the Ottawa Woman's Club when it was organized in 1911. Dr. Pettit belonged to the Grand Army of the Republic lodge.

J. W. and Mary had three children who died in infancy, in 1875, 1879, and 1880.

They also had two sons: Harley Van Alstyne Pettit, born in 1882, and Roswell Talmadge Pettit, born in 1885; the sons spent most of their lives working at the tent colony and T.B. Sanitarium.

Roswell became a physician, graduating from the University of Chicago in 1910 with honors. The university awarded him the Freer Prize, a gold medal, and cash award for the best original research work. He did further medical training in Austria. Roswell was the president and head physician at the tent colony, T.B. Sanitarium, and Illinois Valley Hospital, all on Center Street in Ottawa.

Roswell Pettit was called a hero by the *Ottawa Free Trader* for saving a young woman from drowning on the Fourth of July in 1902. He and six other young men were in a small boat headed to the baseball grounds at Glen Park resort (on the Fox River near Sheridan north of Ottawa) when they saw a small boat with three young ladies and one man capsize. Roswell and the other men in his boat paddled toward the other boat. One young lady drifted from the others. Roswell swam to her and held her head above water until a steamer arrived to pull all of them to safety. The woman was "exhausted and close to drowning," the newspaper account said. Charles Lewis, a clerk at R. C. Jordan's hardware store in Ottawa, was a companion of Roswell who also was cited as a hero. The newspaper account concluded, "For this act of saving four lives, thus avoiding a stigma upon the resort, the Ottawa boys were charged an additional 15 cents for keeping the boat out over time! How's that for Glen Park?"

The house at 823 E. Center Street, which was built for Dr. Pettit.

Dr. Roswell Pettit made a name in his own right. He wrote an article for the November 1911 issue of *Journal of Infectious Diseases* titled, "Secondary Infection in Pulmonary Tuberculosis: The Recovery of the Streptococcus and Pneumococcus from the Blood." He wrote and presented a paper, "The Tuberculin Therapy of Pulmonary Tuberculosis," to the Iowa-Illinois District Medical Association in Rock Island in October 1912. He wrote an article for the May 1924 issue of *Endocrinology*, titled "Differential Diagnosis of Incipient Tuberculosis and Incipient Goiter." He also wrote "Intravenous Pyelography in Renal Tuberculosis" in 1931.

Roswell also was a big fan of the circus. Every summer, he and his wife were guests of Ringling Brothers Circus for several weeks, following the circus on its travels. "There is a fascination and interest in circus life which appeals to the Ottawa physician and he employs this bend in his nature as a means of recreation," according to an article in the *Streator Times* on March 24, 1927. His interest started in 1917 when he met circus owners in Berlin. He gave a detailed talk about circus life to the Streator Rotary Club in 1927.

Roswell married Dorothy Blatchford in 1920 and they had two daughters, Louise More and Frances O'Halloran. Roswell died in 1953; Dorothy died in 1981.

Harley was the superintendent and business manager at the tent colony and tuberculosis sanitarium and his family lived at the facility. Harley married Lucy Belle Nugent. They had a daughter, Nance, who died in childhood in 1911, and three other daughters: Anna (1913–2006), Nancy (1915–1995), and Patricia (1919–2009). Harley died in 1964; Belle died in 1956.

Dr. James Wiley Pettit died on September 3, 1926, at Illinois Valley Hospital in Ottawa at the age of seventy-eight of a cerebral hemorrhage from a stroke in 1925 that had paralyzed him. He spent the last ten months of his life in bed and a wheelchair. The funeral was held in the home of his son, Roswell, at 326 Pearl Street. Rev. H.S. MacKenzie, the pastor of the Congregational Church in Ottawa, officiated. Pallbearers were Al Schoch, James McGrath, Oscar Haeberle, James Herring, Frank Wing, and Charles Woodward.

Mary Louise Pettit died on November 23, 1926, in Illinois Valley Hospital, after a long illness. She was seventy-four years old. The Pettits are buried in Ottawa Avenue Cemetery.

Thomas Rabbitt died of tuberculosis on October 6, 1888. He was born in 1838 in Rochester, New York, the first of eight children of Irish immigrants. Thomas was an iron molder in Rochester before he and his brothers Edward and John enlisted in the Union Army in 1861 after the Civil War started.

John was detailed to the Ambulance Corps, where he rescued wounded and sick soldiers, pulling them into his horse-drawn ambulance cart and bringing them to doctors. Disease was rampant, and tuberculosis killed approximately 14,000 soldiers during the war.

After the war, Thomas headed to Ottawa, Illinois, where he met and married Ellen Duffy. Ellen's father, Patrick Duffy, an Irish immigrant from Galway, settled in Ottawa in 1840 and worked on the Illinois–Michigan Canal, earning enough to purchase a 120-acre farm by 1877.

Thomas and Ellen moved to St. Louis where Thomas worked as an iron molder. They returned to Ottawa in the early 1870s to raise their family of two daughters and four sons.

In the 1880s, Thomas knew something terrible was happening in his body. After several years of experiencing all the symptoms, tuberculosis was diagnosed.

Sharon Page, the wife of Thomas Rabbitt's great-great grandson, remarked:

The stigma of tuberculosis was a psychological burden for the infected person. Everyone feared being infected, so the community would shun families with the disease. Thomas knew there was no cure for him. His fear of the disease may have been heightened with the death of his brother John. Thomas also knew he needed to protect his six children. As Thomas' symptoms worsened, he made the decision to separate himself from his family towards the end. He knew he would never see his family again when he left his home in the fall of 1888. He was a selfless man.

Thomas took the train to Clinton, Iowa, 100 miles from Ottawa. When he arrived in Clinton, he rented a room in a house at 216 First Avenue. He was already in the final stages of tuberculosis with weakness, extreme weight loss, lack of appetite, fever, chills, and coughing up blood or phlegm. Thomas survived only two months.

Dr. James Keho described Thomas' death on October 6, 1888, as *Phthisis Florida*, an acute, rapidly fatal pulmonary consumption, also called "galloping consumption." His body was buried the same day in St. Boniface Catholic Cemetery in Clinton. He was fifty years old. Nine years later, his brother, Edward died of tuberculosis, at the age of fifty-seven; his other brother, John had died of tuberculosis in 1883 at the age of forty-one. Ellen died in 1898.

Thomas' oldest son, Thomas Edward Rabbitt, Sr., was seventeen when his father died. During the next decade, he continued to live with his widowed mother and became a glassblower for the La Bastie Glass Works in Ottawa. He married Lizzie Kenney in 1897 in Ottawa. Sharon Page related:

He gave her a wedding gift of a hand-blown amber and clear glass garland with a heart in the middle to drape over their side doors. Two children followed: Ellen in 1898 and Thomas, Jr., in 1900, my husband's grandfather.

In 1903, Thomas and Lizzie were expecting their third child. Suddenly, in late August, Lizzie suffered for seven days with acute uremia at Ryburn Memorial Hospital in Ottawa. She and their baby died, and the Rabbitt and Kenney families buried mother and child together in Saint Columba Cemetery in Ottawa. When the Kenney family returned home to Nebraska, they took the two other children with them; Thomas, Jr., lived with an aunt in Lincoln, and Ella lived with her grandmother in Milford. Thomas, Jr., saw his father only once when he was older.

Thomas Edward Rabbitt, Sr., died in 1946. His daughter Ella died in 1990 and his son, Thomas, Jr., died in 1972.

Sharon Page commented:

The Ottawa Tent Colony opened and Thomas Edward Rabbitt Sr. started to work there. The pain of losing his father to tuberculosis must have guided this decision. In a photo postcard sent to his children, he proudly wrote, "Bringing hot dinners to the tents for the sick people. Papa." It was more than a job for him; it was an emotional connection in helping others.

Above: Thomas Edward Rabbit, Sr., on the far right, is helping deliver meals to patients in the Ottawa Tent Colony.

Left: Thomas Edward Rabbitt, Sr.

Judy Corcoran was eleven years old in 1957 when she contracted tuberculosis.

> I contracted T.B. when we were renting the old house on Gentleman Road across from the stables. I used to play with a family that lived behind the stables. It was a huge house. I was like an only child—my brother Wayne was 18 years older than me—so I loved going to that house full of people.
>   One of the men who lived there was an active carrier, so it is assumed that is where I contacted it.

Judy was attending the Fall River Township rural school at the time. "There was no home schooling back in my T.B. time. My teacher was so helpful stopping to see me with some school work. She passed me on to seventh grade, but I had missed April, May and some June."

Judy was twelve when she was admitted to the LaSalle County T.B. Sanitarium, across from the former tent colony. "I'm sure glad it was a building by then."

"My mother told me I was just going for a couple of days. When I was admitted, a nurse told us they didn't wash the patients' hair for two weeks. I looked at my mother, thinking, 'You said two days.'" She ended up staying for five months. Judy was the youngest person in the institution.

Above left: Judy Corcoran holding her niece Cindy at Christmas 1957, after leaving the Ottawa T.B. sanitarium.

Above right: A sanitarium newsletter, a souvenir from Judy's stay.

Her mother, Margaret, visited every day and she was allowed to come right into Judy's room, but there were restrictions. When her brother Wayne's wife had a baby, Judy could only see her little niece, Cindy, from her second-floor window. "My brother would come and always bring me a chocolate milk shake. The idea was to fatten me up."

Judy remembers moving her meal tray to the doorway of her room so she could eat with the patient in the doorway across the hall. On a bathroom break, she would sneak down to the solarium where other patients were watching television or playing cards.

I would wait for a nurse named Sophie to walk down the hall every Friday, hoping for her to stop in my room. What she handed out was called "privileges," starting at being able to go to the actual bathroom and eventually full bathroom trips. Oh, how we waited to hear her footsteps and hopefully enter our room with the new privilege. I quickly figured out they did not keep track of when you went, or at least I thought they did not. The nurses would change in the afternoon so I was using that bathroom more than I was supposed to. There was a solarium at the end of each hall and you were also able to go there so many times a day. I also got the privilege of going home on a Sunday twice near the end of my stay.

When I first got there, I was on the first floor of that yellow brick building and shared it with Helen Poldek from Streator. I was not a Catholic at the time, but Helen would come to my doorway every night to recite the rosary with me.

I remember standing on a chair in the corner of my second floor room watching out the window being able to see some fireworks on the Fourth. To this day, I never try to miss watching fireworks.

Judy said she got a gift every day for a month from the Paul Street Bible Church, where she had attended. She also still has the beautiful handmade card from her classmates at Fall River Township School. It was a lonely existence without her family, and there were no phones in the room.

To this day, I try to send cards to anyone in the hospital or sick at home, remembering how much it meant to me getting mail.

When I left the sanitarium, we had moved to Ottawa Avenue, renting an upstairs apartment. The doctors warned me that some people might shun me. So the next day after I was released, I headed to a friend's house, Jill Anderson. I called Jill and out she came. I was so happy. Jill and I have remained close friends to this day.

Judy met Bob Dubach when she was a freshman and he was a senior at Ottawa Township High School. They married after she graduated in 1963. They have four children: Debbie Inocencio, now in California; Darci Aubry in Utica, Illinois; David in Ottawa; and Bobby in New York. Bob and Judy have five grandchildren and one great-granddaughter. Bob worked for AT&T. He and Judy moved from Ottawa to Georgia in 1990.

TB has only affected my life is that now I have rheumatoid arthritis and there are certain drugs I cannot take, being I have had TB. The only effects I remember from TB was I lost weight and am more tired than usual. But after my five-month stay, I felt fine.

Kathalyn Anne Holz-Zammuto of Ottawa said:

> My husband's mother was a patient for six months in one of these sanitariums when he was five years old. It was very traumatic for him, not understanding where she was and not being able to visit. She went to one in Arizona.

Diane Whalen of Ottawa said:

> My mother, Elizabeth June Cunningham, spent nearly one year in the TB sanitarium and passed away in 1963 when I was nine. I remember going there on Sundays and talking up into her window. Only my Dad was allowed inside. I remember her getting certain 'privileges,' into the hallway, into social rooms, outside, etc. She was allowed to leave for her Father's funeral, but us kids could not go near her. It is a sad memory of childhood.

An article in *The Daily Times* in 1996 by Peggy Schneider told the stories of James Milus, Wayne Erickson, Pluma Groesbeck, and Marilyn Renz. While his friends were entering the military in 1940 as World War II approached, Mr. Erickson entered the LaSalle County T.B. sanitarium. He was nineteen. He had wanted to join the Army Air Corps and fly airplanes, but T.B. ended that dream. He eventually saw action in the Merchant Marine. Mr. Erickson remembered a hospital routine where patients were awakened at 6 a.m., bathed, served breakfast, and then given rest periods in the morning and afternoon, with lights and radios off. He said the rich diet caused him to have a heart bypass operation later in life.

Mr. Milus, a member of a bomb disposal unit in North Africa during World War II, had to give up his job in a carton manufacturing plant in 1947 to enter the county T.B. Sanitarium. He was newly married at the time. For Christmas, his friends sent him a money tree. He said he played 4,000 games of two-handed pinochle in the sanitarium, and also made rugs and doilies.

Mrs. Renz had a three-month-old daughter and two-year-old son when she entered the sanitarium in 1936 at the age of twenty. She underwent surgery to have a part of her lung removed.

Mrs. Groesbeck learned to crochet and she wrote poetry to fill the time. Remembering the feeling of confinement, Mrs. Groesbeck told the reporter, "I can sympathize with anyone who is in prison." Mrs. Groesbeck became pregnant again while recuperating; she gave birth in Ryburn Memorial Hospital, and her baby stayed in the hospital nursery for several months while Mrs. Groesbeck went back to the sanitarium.

Assistant administrator Ruby Jovanovich told Peggy Schneider that she recalled looking out the window and seeing the foundation where the barracks had been before the new yellow brick addition was built. She also recalled all the windows being open, and the strong odor of ointment used on sore backs and other aches. Some items were smuggled into the building, including bottles of liquor pulled into second-story windows on a string.

Emma Kristina Pierson was born in Sweden in 1880 and came to America through Ellis Island, along with her brother, August. She worked as a housekeeper in Joliet before coming to LaSalle County. She contracted tuberculosis and spent time in the

Emma Pierson Dumke.

Ottawa Tuberculosis Sanitarium. She married Henry Dumke (1889–1967) and they had three sons: Leonard Dumke (1914–1987), Robert Dumke (1915–2001), and Frederick Dumke (1925–2009).

Henry Dumke owned a greenhouse in Marseilles for more than fifty years. His son, Robert, owned a greenhouse in Ottawa for decades. Emma died in 1932 at the age of fifty-two. Her older sons had to leave high school to take care of the youngest brother.

Agnes Magoonaugh was born on April 20, 1900, to Thomas and Velma Magoonaugh. Agnes had a brother, James O'Kelly Magoonaugh (known as Kelly), and a sister, Thelma. Agnes contracted tuberculosis in 1918 when she was a junior at Ottawa Township High School. She lived at 1245 Phelps Street. She died on September 28, 1927, at the LaSalle County Sanitarium, where she had been a patient for eight years. She lived most of her time there in a 9 × 10 wooden hut.

Her obituary said, "She was extremely popular with her classmates. Before her illness, Miss Magoonaugh was prominent in St. Francis Catholic Church circles and had an active part in the dramatic society."

The funeral for Agnes Magoonaugh was at St. Francis Catholic Church in Ottawa, with William Schomas, Cecil Dougherty, William Zwanzig, William Hossack, Henry Schiffgens, and Oren Rising as pallbearers. She is buried in Calvary Cemetery, Ottawa.

Thelma Magoonaugh and Velma Bounds were in the tent colony at the same time. They both tested positive for T.B., but they did not have it.

Right: Agnes Magoonaugh and her sister Thelma (Magoonaugh) Frizol.

Below: Agnes Magoonaugh, James O'Kelly Magoonaugh, Velma (Bounds) Magoonaugh, Thelma (Magoonaugh) Frizol.

Left: David Braham, a patient in the Ottawa Tent Colony in 1913.

Below left: Helen Voltmer.

Below right: Frances Bodley.

Kelly Magoonaugh, Sr., met Velma at the tent colony, when he went there to visit his sisters, Agnes and Thelma. Kelly and Velma started dating when she got out of the clinic. They married in 1933.

They had a son, James O'Kelly Magoonaugh, Jr. (also known as Kelly), born in 1938. They lived at 1324 W. Madison Street. Velma Magoonaugh later worked at the sanitarium as a nurse. Velma died in 1984. Kelly, Sr., died in 1975.

Helen Voltmer graduated from Lakeside Nursing School in Chicago in 1917. In her class was her lifelong friend, Frances Ann Bodley Maley, who was from the Fairbury area.

Helen was superintendent of nurses at the Ottawa sanitarium from 1935 to 1949. She went to New Mexico in 1949 and continued her career as a nurse. She died there in 1976. She never married.

Tuberculosis was so prevalent in everyday life that it became part of the culture. It was included in many novels such as Anton Chekhov's *Anna Karenina*, Charles Dickens' *Nicholas Nickleby*, and Victor Hugo's *Les Miserables* and operas such as Verdi's *La Traviata* and Puccini's *La Bohème*.

The T.B. pandemic of the 1800s greatly influenced Victorian fashion. The pale skin, thin figure, rosy cheeks, and lips from a low-grade fever were all thought to be attractive qualities in upper-class women, according to an article in *Smithsonian Magazine* in 2016. Corsets and makeup also highlighted what was known as "consumptive chic," the article said. Women's hemlines became shorter because long dresses that touched the floor could pick up tuberculosis germs from spit on the ground. Men's fashions also changed, as beards and mustaches, which were believed to harbor tuberculosis and other germs, went out of style.

Hygiene was taught in schools to prevent the spread of infection. Public parks, with open space and playground equipment, became more popular to encourage fresh air and exercise. Even the creation of the ice cream cone was to promote sanitary actions. Ice cream was served in small glass dishes, which customers would lick clean, and then be reused by the merchant; the edible cone was invented to stop the spread of the germs. The cone was invented in 1903 and became firmly established at the 1904 World's Fair in St. Louis.

Thousands of famous people were afflicted. A few of those who died from the disease include: novelists Charlotte and Emily Bronte and their brother, Branwell; poets Elizabeth Barrett Browning, Robert Burns, John Keats, Dylan Thomas, and Walt Whitman; writers Henry David Thoreau, Anton Chekhov, Franz Kafka, D. H. Lawrence, Robert Heinlein, Washington Irving, Stephen Crane, Thomas Wolfe, and George Orwell; composer Frederic Chopin; Sir Walter Scott; Voltaire; Honore de Balzac; actress Vivian Leigh; playwright Eugene O'Neill; John Henry "Doc" Holliday, famous gambler and gunslinger; South American hero Simón Bolívar; King Louis XIII and King Louis XVII of France; King Edward VI of England; Cardinal Richelieu of France; and Saint Bernadette of Lourdes. It is believed St. Francis of Assisi died of tuberculosis in 1226 at the age of forty-four. Egyptian Pharaoh Akhenaten and his wife, Nefertiti, both died from tuberculosis.

Abraham Lincoln's sons, Thomas "Tad" Lincoln, died at age eighteen, and Eddie Lincoln died a month before his fourth birthday. Jimmie Rodgers, one of the greatest figures in country music history, sang about his struggle with T.B. before it killed him in 1933 at the age of thirty-six.

5

Dueling Newspapers

There always are wars between rival newspapers. In the nineteenth and twentieth centuries, these wars often were bitter and nasty. Many times, they were fought over differing views on politics, business, or other subjects. *The Ottawa Free Trader* was supportive of the tent colony; *The Ottawa Journal* was not.

There are not any surviving issues on microfilm of the *Journal*, but we do have access to *The Ottawa Free Trader* for its comments on *Journal* stories. This was in the December 15, 1905, *Free Trader*:

In view of the high encomiums passed upon the Tent Colony in charge of Drs. Pettit and Butterfield of this city by the hundreds of leading physicians who paid it a visit yesterday, one would think the publishers of the *Journal* of this city should be ready to hide their heads in shame and confession over the coarse, mendacious and wholly unjustified attack on that institution in their last Sunday's issue. That the attack could be prompted only by a mean spirit of jealousy and spite is evident from the gross falsehood of all its main statements. It is not true, for example, that the colony or "camp" is located "within this city," for as we all know, its location is a mile and a half to two miles from the main city and on the opposite side of the Illinois River.

Then, in the second place, it is equally untrue that patients of the colony are frequently seen on our streets, threatening to spread the "white contagion" among all with whom they come in contact. Nor, lastly, has the *Journal* given any reason to believe that tuberculosis is a contagious disease at all. The highest medical authority of the world has long ago settled this question beyond all controversy. It can be communicated, the better belief appears to be, by inoculation with the sputum of an advanced consumptive, but not by simply breathing the same air or coming in contact with a patient.

On this head we can give proof from personal observation that is conclusive. We recall a family of 13 children with which we had close connection, the father of which lay for a year with consumption on a bed located in the common living room of the family where he was daily surrounded by all of them and all breathed the same air with him. Yet of this entire family, not one ever took the disease. None died under the age of 60 and one at least attained the age of 90.

Instead of being a menace, the Tent Colony is a credit to Ottawa, an institution of which the city should be and is proud, giving it prestige abroad as having located in its vicinity a refuge of sure hope to victims of a plague which, up to a few decades ago, was dreaded as equal in fatality to the "black death" that ravaged Asia and Europe in the Middle ages.

The Free Trader excoriated the *Journal* in an earlier editorial on January 20, 1905:

Suffering from a severe attack of the collywombles, it has had a nightmare in which it has seen terrible things growing out of the establishment of this colony. And this was not unexpected, neither the attack nor the source from which it emanated.

It accused the *Journal* editors of "phthistophobia ... an unfounded and often hysterical fear of contracting consumption, and it has been a clog on the anti-tuberculosis crusade along many lines, besides causing untold and needless suffering."

The fact cannot be too often repeated or too insistently pressed, that a consumptive need never be an object of dread to an intelligent man, or even to a stupid one, when the patient himself is intelligent and careful. For it is really a difficult matter to catch tuberculosis except by sleeping in an infected room, eating with infected ware or working for a considerable period in a place where consumptives have been spitting.

A more perfectly developed or more acute "phthistophobia" has probably never been seen than that exhibited by our esteemed contemporary. It sees visions of Ottawa with its streets deserted and its trade destroyed as a result of this "terrible invasion" that exists only in its perfervid imagination. The one concrete statement on which it bases its otherwise inane twaddle is the fact that some barber has objected to shaving some patient from the colony.

This outburst by the local sheet is the regular "last stand" that is always made by the recalcitrant forces of every community in the face of progress. It is exceptional only in the baselessness of its virulency. So it is probable that the work of the Tent Colony will go on without any general uprising of the people. It seems to be doing good work and to be ideally located. So far there has developed but one drawback to its location here—and that the publication of such pessimistic and uncoalescent wails in its immediate vicinity. Should the state propose to locate near here another great hospital for the treatment of the insane, it would be perfectly consistent for the same publication to oppose its establishment on the ground that its presence would be dangerous to its staff because of their peculiar susceptibility to contract disabilities of that kind.

The article concluded with an anecdote from Samuel Hopkins Adams in *McClure's Magazine*:

The client had a friend who had developed consumption and he practically besought the doctor to advise him to keep away from his friend. The doctor replied, "Certainly keep away from him. It wouldn't hurt you to be with him, but it would probably depress him to have a sniveling coward like you about."

The *Ottawa Journal* was published from 1898 to 1916, when the two newspapers merged and became the *Free Trader-Journal,* and later the *Ottawa Fair Dealer.* It ceased in 1927. The more established *Daily Republican-News* survived, and it eventually became *The Daily Times* and then *The Times.*

The editor of the *Marseilles Plaindealer* also derided the tent colony's mission, even before it opened, and with statements that were not true. The *Free Trader* quoted some of the *Plaindealer's* story in a June 24, 1904 edition:

> In spite of our warning to the people and the press of Ottawa, they appear to glory in the thought that a consumptive cure is to be established at Ellis Park, and what is more they are trying to prove that the disease is not of the transmitting kind. None are so blind as those who will not see, and this way of repudiating our advice will surely bring retribution upon their heads.

The *Free Trader* editor then spent a considerable amount of ink letting the Marseilles editor know how wrong he was, citing statistics from the state medical society—6,895 tuberculosis deaths in Illinois in 1902 and 7,026 deaths in 1903. The average age at death was thirty-four.

Aristocracy Row, Tent Colony for Cure of Tuberculosis, Ottawa, Ill.

The Club House, and a close-up of patients and cottages.

First Congregational Church, Ottawa (left), and First Presbyterian Church, Ottawa (right).

Washington Park in Ottawa, where Abraham Lincoln and Stephen Douglas debated in 1858. St. Columba Catholic Church and Reddick Mansion are in the background.

The LaSalle County courthouse and LaSalle Street in Ottawa, in the era of the tent colony.

Ottawa, as seen from across the river on the south side.

The Hilliard Bridge across the Illinois River, connecting Ottawa's north and south sides.

6

BUFFALO ROCK TENT VILLA

In addition to the Ottawa Tent Colony and the LaSalle County Tuberculosis Sanitarium, there was another local tuberculosis facility—the Buffalo Rock Tent Villa Company.

Buffalo Rock has an interesting history prior to becoming a tent colony in 1908. According to legend, Buffalo Rock got its name because Indians would drive buffalo to the top and then force them over the edge of the bluff, which would break their legs and make them easier to capture. Buffalo Rock was the home of the Illinois Indians when explorers Louis Joliet and Father Jacques Marquette made their historic trip up the Illinois River in 1673. In August 1680, the Illinois tribe was almost wiped out in a war with the Iroquois Indians.

Buffalo Rock was a French military, missionary, and trading post called Fort Miami. During the winter of 1682–1683, explorers Robert Cavelier deLaSalle and Henri deTonti built Fort St. Louis on nearby Starved Rock and gathered 4,000 Indian warriors at Buffalo Rock to form a confederation against the Iroquois. The Miami Indians was one of the tribes in the confederation. The Illinois, Miami, and Iroquois Indian tribes had camps on Buffalo Rock at various times.

Across the river to the west is Starved Rock. According to legend, the Potawatomi and Fox tribes attacked the village of the Illinois (Illiniwek) Indians by the Great Rock in 1769 to avenge the death of Chief Pontiac. The Illinois tribe retreated to the top of the rock. They were trapped without food or water and starved to death, which gave it the name Starved Rock. Some historians question whether this tale is true.

Buffalo Rock and Starved Rock were owned by private citizens. Daniel Hitt bought Starved Rock from the federal government for $85 in 1835. He sold it in 1890 to Ferdinand Walther for $21,000. Walther developed it into a recreation area. The state of Illinois bought Starved Rock for $146,000 in 1911 and made it a state park, which now encompasses more than 2,600 acres. Nearby, Frederick Matthiessen's property, known as Deer Park, was donated to the state following his death in 1918. Matthiessen State Park now totals 1,938 acres, with stunning canyons, bluffs, and waterfalls.

Buffalo Rock was owned by Lon Edwards in the 1890s. It was purchased in 1897 by Duke Farson, a millionaire banker and bond broker turned Methodist preacher. He bought the rock and spent $5,000 improving the grounds before he began his religious revival camp meetings there.

BUFFALOES AT BUFFALO ROCK STATE PARK, OTTAWA, ILLINOIS

ILLINOIS-MICHIGAN CANAL, BUFFALO ROCK SANITARIUM, OTTAWA, ILL.

Farson, whose name is still well known today, was pastor of the Metropolitan Holiness Church in Chicago. It was Farson and fellow Chicago businessman E. L. Harvey who founded the Metropolitan Church Association in the early 1890s. William Kostlevy, in his 2010 book *Holy Jumpers*, wrote:

It was one of the most controversial societies of the era. Its members were called "jumpers" because of their acrobatic worship style, and "Burning Bushers" after their caustic periodical, *The Burning Bush*. They objected to the concept of private property, rejected "elite" denominations, and professed an alternative, radical vision of Christianity, using modern music and folk art to spread their message.

Farson and Harvey created the Burning Bush Movement in the 1890s within the Methodist-Episcopal church. They left the Methodists and joined the Holiness Movement in 1899. They stressed communalism and socialism combined with religion.

An account from old records of his church tells how Farson came to buy Buffalo Rock:

There is an interesting story in regard to Mr. Farson's purchase of the place. He was delayed through missing a train, and took a hack to go to another place. When he passed the spot, he was immediately attracted by its beauty. Three men had just offered the owner the sum of $9,000 to purchase the grounds for a beer garden. Mr. Farson paid the sum of $10,000 to use it for religious purposes.

There was no admission fee for his camp meetings and dinner was provided for a free-will offering. One meeting in September 1902 made a $2,000 profit, with another $100,000 donated.

The locals did not like Mr. Farson. An article in the June 13, 1902 *Ottawa Free Trader* told of a revival meeting planned for August. Buffalo Rock had numerous signs warning trespassers to keep out, with fences and men to enforce the warnings. Citing details from *The Burning Bush* (which the editor called *The Hot Shrub*), the article said 1,500 trees had been planted; a tabernacle and a barn were being built; and the slough was being transformed into a lake. The entire 102 acres had been deeded to the Camp Meeting Association.

The editor of the *Ottawa Free Trader* wrote this column in the June 20, 1902, edition:

The ideas of hospitality of the holiness sect, as exemplified by Mr. Duke M. Farson, proprietor of Buffalo Rock, as well as the conception of the object the Almighty had in lavishing on LaSalle County the entrancing scenery along the Illinois, Fox and Vermillion rivers, are certainly original and hardly, we think, in strict accordance with the scriptures they profess to live so closely by. Buffalo Rock, particularly the eastern end, has from the days of the earliest Injun who wandered through the Illinois valley been recognized as one of the beauty spots of the country, the view from the summit of which makes any man the better for having beheld it. Access to the rock at this point has always been free as air and the view has delighted thousands and brought them a little nearer the Maker of it all.

But things have changed since the banker evangelist has secured possession. If you haven't got religion, you needn't come 'round. The place is placarded with signs of "No Admittance," "Keep Out," "This Is Private Property—No Trespassing," etc., and the saved keeper, in accordance with his instructions, sees that the mandates are obeyed.

One hot day last summer, a couple of canoeists who were enjoying a brief respite from their labor in a close and sultry shop by taking a day with nature along the Illinois, stopped at Mr. Farson's well for a drink of water, and lying on the grass nearby were enjoying the beautiful scene before them . While they were drinking in the scene—they had fortunately finished drinking in all the water they wanted—Mr. Farson's redeemed tenant approached and demanded, "Don't you men know you are trespassing on Mr. Farson's private property? You have no right to get water here and will have to get off this property at once."

The unsanctified canoeists were somewhat astonished, but on recovering and preparing to make their departure said to the redeemed servant of the Lord's anointed Mr. Farson, "There is no use, of course, arguing with you, as you are simply obeying your master's orders. But you tell the Rev. Duke M. Farson for me that at least one poor miserable sinner thanks God that he owns but one beauty spot along this magnificent river and can deprive even the least of these of but one view of this paradise."

A story in October 1902 told of a young woman who started screaming and fighting with her mother and with police during a Holiness meeting at Farson's Metropolitan Methodist-Episcopal Church in Chicago. And a seventeen-year-old girl refused to go home with her mother after a service, and also fought with her mother and the police. Several hundred followers of Farson blocked the street in front of the police station, demanding the girl's release. The *Ottawa Free Trader* reported, "Several persons living near the Metropolitan Methodist-Episcopal Church say that it is a regular occurrence every time a revival meeting is held to see members of the congregation leave the church apparently bereft of reason. Many families refuse on this account to permit their children to attend the services."

The footbridge at Buffalo Rock.

After a Holiness revival meeting in October 1902, *The Rockford Republic* printed:

Duke Farson and (evangelist) Billy Sunday entertain the same opinion concerning Rockford. Some time ago, when the ex-ball player was stirring up animals in Belvidere, he told his audience that Rockford was hell, and if they wanted to go to hell all that they had to do to get there was to take a car to Rockford. Last night at the Holiness meeting at the opera house, Duke Farson told his large audience that Rockford was hell and nearly all of the people were disciples of the devil.

The *Rockford Republic* printed on August 7, 1903:

Duke Farson, the real shouter of the Holiness Church, and the man who seems to actually believe that should be a part of everyone's religion, is to be the leader tonight at the Holiness camp meeting at Harlem Park. Farson left the Methodist Church, in which he was a lay preacher, because the Methodists did not shout enough, did not show joy by dancing, etc., and did not preach enough about Hell to suit him. Now he has an independent church of which he is the principal moving force, he does all that he thought the Methodists ought to do.

The Rockford newspaper reported on August 8, 1903:

Tales of how converts to the Holiness faith had visions, visited Heaven, were healed miraculously, were cleansed of devils and many other things that sounded strange and uncanny were the principal part of a sermon delivered last night by Rev. Duke Farson before an immense congregation at the Harlem Park auditorium. The spirit once more worked in the Holiness camp meeting, for Duke Farson seems to have a peculiar faculty of making it move. There was jumping, shouting, moaning and groaning, laughing, dancing, singing and yelling at the meeting last night, and despite it all the audience kept quiet and listened to the noise that was made by the workers and their converts.

The newspaper also told details about Farson healing the sick, people seeing Heaven in a trance, and more.

In August 1904, Rev. Charles Wetherell, head of the Rockford Holiness Movement, quit the church and brought the entire congregation with him because he thought Farson was "hysterical" and a "fanatic."

Arthur Bray, a follower of Farson, also became an important religious figure in the twentieth century. Mr. Bray left the Holiness church in 1937 and attended Olivet Nazarene College in Bourbonnais (which was founded in 1905 as Illinois Holiness University). He became a Nazarene minister before becoming a Methodist minister.

Duke Farson—whose birth name was Marmaduke Mendenhall Farson—moved to Los Angeles, California, in 1927 to be a pastor of a Holiness church. He died there in 1929 at the age of sixty-five.

Mr. Farson put Buffalo Rock on the market in June 1904. The Ottawa *Daily Republican Times* on June 17 observed that it could be sold to a company that would grind it down for glass sand. That whole area of Ottawa and west to Buffalo Rock is rich in silica sand. Ottawa Silica Company has been in business for more than 100 years,

with huge, deep quarries where mining is still being done. The newspaper urged some enterprising men to buy the rock and turn it into a recreation area. Several years later, after Buffalo Rock had been sold and the tent colony came and went, it was put back on the market and the sand companies once again tried to acquire it.

It was early in 1907 when plans were announced to start a tent colony for T.B. patients there. Buffalo Rock Tent Villa opened on May 1, 1908. The facility only accepted patients in the early stages of tuberculosis. Its opening capacity was fifteen patients. Rates were $25 to $40 per week. The resident physician was Dr. E. Don Taylor of Chicago.

The first patients arrived on May 11, 1908. Dr. Joseph Cobb, a prominent Chicago homeopathic physician, was the chief organizer. Dr. Taylor ran the facility. A brochure read:

> The Buffalo Rock Tent Villa treats all kinds of chronic diseases and makes special provision in separate tents for tuberculosis patients. The Colorado house tent is the unit of the Villa. Cottages and large screened porches are also used. The sanatorium is situated on a rocky island, 60 acres in area, three miles from any town or city. Application should be made to the resident physician.

An advertisement called it "A Sanitarium for the care of All Kinds of Chronic Diseases, including Neurasthenia, Mild Dementia, Convalescents from Operations and Acute Diseases and Drug Habitués. Tubercular Patients in the incipiency of the disease will be accepted." Dr. C. A. Laffoon was listed as the resident physician and assistant superintendent.

In addition to tubercular patients, it was open to others who needed "rest, diet, fresh air and sunshine."

Tents in Construction on Buffalo Rock, Colony for Cure of Tuberculosis, Ottawa, Ill.

Main Building at Buffalo Tent Colony Near Ottawa, Ill.

Spring water was pumped in. One building was all glass on all four sides. The facility was described as buildings with concrete floors, ample kitchens, dining rooms, and sun parlors. Later, an administration building and an amusement hall were built. The large amphitheater built by Farson was still there, and there were a number of tents in place.

Twenty homeopathic physicians, on their way home from the annual meeting of the state medical society in Chicago, stopped by Buffalo Rock for the dedication of its tent colony on May 16, 1908. A number of other guests were there and were treated to a nice luncheon. The *Ottawa Free Trader* reported, "The visitors examined the new rest cure and were loud in their praises of its ideal location and its fine equipment. The visitors got to the rock under trying circumstances. For two or three hundred feet at the base of the rock, the roadway is under two or three feet of water of the Illinois. The visitors were ferried over in a wagon. After their visit, they were taken for a further trip down the line in the special interurban car. Dr. Taylor was the host. Among the guests was his father, who built the state asylum for the insane in Watertown and was superintendent."

In November 1909, Dr. E. H. Butterfield became medical director at Buffalo Rock. He had been associate director of the Ottawa Tent Colony with Dr. Pettit.

The Buffalo Rock Tent Villa did not fare well and it closed after a couple of years. Ownership of Buffalo Rock went to J. M. Hopkins of Chicago, one of the stockholders. The owner put the rock up for sale in 1911, including the buildings designed for patients.

It was feared in 1911 that Buffalo Rock would disappear. "Some of these days Buffalo Rock may be blotted off the face of the earth," according to a story in the *Ottawa Free Trader* on September 1, 1911. "And no earthquake is needed to do the work at that. There are a number of the big sand companies that have their eyes upon it. Once let them get possession of it and it would disappear in the course of a few years. And this would be a distinct loss to the county and the state."

The newspaper proposed that the county buy the rock and the farmland around it. The writer noted that the state legislature recently authorized counties to establish T.B. sanitariums and suggested that Buffalo Rock would be a suitable site. It would be a public sanitarium, unlike the privately-owned Ottawa Tent Colony. Dr. Pettit was in favor of it.

The Woman's Club of Ottawa spoke to a large crowd in the high school on May 16, 1912, about saving Buffalo Rock. The club passed a resolution, lamenting that the county board of supervisors declined to purchase the rock, stating that the owner's "public spirit is at so low an ebb that measures cannot be devised for the preservation of this property." It resolved to form a committee to find a way to preserve it. Two weeks later, the women announced that Mr. Hopkins agreed to postpone the June 14 sale of the property. The women planned a third appeal to the county board (or to the township board) to buy Buffalo Rock for a recreation park. "The women have determined to save Buffalo Rock and when they set out to do a thing they generally do it," the *Ottawa Free Trader* said. By October, the women had raised $7,000 in pledges.

There was a lot of support from civic groups for the LaSalle County Board of Supervisors to buy Buffalo Rock, particularly since a county tuberculosis sanitarium was needed and there already were buildings there for that purpose. Yet the county board just would not do it. The *Ottawa Free Trader* was among those pushing for it, criticizing the county board for its inaction.

The headline in the October 27, 1911, edition read, "No Money To Save Lives Seems To Be The Position Of Many Eminent Statesmen." A committee of county board members went to inspect the property and the buildings. The owners wanted $25,000. That included 95 acres of land, of which was 40 acres on the rock with the rest as farmland. They figured the sanitarium buildings were worth $15,870 of that amount. The land was located next to the county poor farm. Board members were divided on the issue. One said he was opposed to the county running a sanitarium; another said there was just one tubercular patient in the county home and that patient belonged in the pest house. The newspaper said the county board could have bought the rock a few years earlier when they had a chance to get it at a good price, and it chided county board members for going out to inspect the property just to collect *per diem* pay without a serious intention. "The board demonstrated once more that it is possessed of no more foresight than a china nest egg."

The *Ottawa Free Trader* continued to urge the county to buy Buffalo Rock. In a December 1911 edition, it said a public sanitarium was needed because there were a large number of people in the county with tuberculosis, with the poor especially needing the help:

If the cows at the county farm were threatened with the spread of tuberculosis or if the hogs down there were threatened with an outbreak of cholera, the county board would soon get into action. Veterinarians would be summoned and every agency of modern cure and prevention would be invoked.

Indeed, tuberculosis in cattle in Illinois was reported several times in those years. One report from December 1908 from the University of Illinois called the problem "serious and widespread." Hog cholera always is a problem.

The editors were beside themselves in a story in May 1912. The headline read, "Buffalo Rock Is To Disappear Shortly." Smaller headlines read, "Sand Company Will Grind It Up/Might Have Saved It If The County Board Had Been Alive To What Were The Real Best Interests Of This County." The lead paragraph read:

Buffalo Rock, which has figured in romance and history ever since the seventeenth century, is to be sacrificed to the greed of commercialism and is to be torn down and shipped away by the sand operators.

The story went on:

The Carolina poplars skirting the drive have been cut down and the road has been plowed up and contracts have been drawn for the transfer of the Rock itself to a sand company. A reservation has been made of strip along the east side of the Rock for a switch track. The buildings on top of the Rock will be removed and in time the entire Rock will be demolished and shipped to the foundries in Chicago and Indiana for molding purposes.

In May 1912, Mr. Hopkins offered it to the county at a lower price than the sand company offered. LaSalle County was not interested. Mr. Hopkins later sold it to Robert O'Meara.

View on Illinois River from Buffalo Rock Tent Colony, Near Ottawa, Ill.

In 1912, the Richard Crane Company bought Buffalo Rock. Crane employed about 5,000 men and women manufacturing pipe fittings and brass items. The company said it would use the site as a recreation and vacation area for its employees, and it would refurbish and expand the sanitarium buildings as a medical facility for its sick and disabled employees. The *Ottawa Free Trader* wrote:

> That Buffalo Rock did not fall into the hands of the sand pirates is due to the diligence of the Ottawa Park Commission, comprised of Dean Quinn, Horace Hull and Oakley Esmond, and the backing they were given by the city's women hustlers, who were prone to see the site destroyed by blasting powder and the steam shovel. J.M. Hopkins, into whose hands the Rock fell as one of the mortgagees after the Buffalo Rock Tent Villa Company has failed, was also watchful against the sand merchants, and only as a last resort would he have sold to them.

The Crane Corporation donated Buffalo Rock to the state in 1928.

Today, Buffalo Rock State Park is 298 acres across the Illinois River from Starved Rock State Park. The park, which has a nature preserve and is a popular picnic spot, is 85 miles southwest of Chicago. Its sandstone bluffs were carved by the Illinois River near the end of the Pleistocene epoch ice age. A few buffalo live in the park as a tourist attraction. There are two trails: the River Bluff Trail, which runs above the Illinois River and has two observation decks, and the Woodland Trail, which is lower in the park with close views of plants and wildlife. Camping is allowed, and there are shelters and stone fireplaces for picnics.

Portion of Driveway, on Way Up Buffalo Rock to Tent Colony, Near Ottawa, Ill.

7

The LaSalle County Tuberculosis Sanitarium

A sanitarium was the goal of Dr. Pettit, even when he began the Ottawa Tent Colony. The tent colony was a beginning—a way to prove his belief that fresh air, rest, and diet were the needed therapy, along with the idea that climate did not matter.

It was hoped that Buffalo Rock might be the site of a sanitarium. It already had the buildings from another tent colony, and it was next to the county poor farm.

Voters in LaSalle County approved a tax to build a T.B. sanitarium; enough tax money was collected by 1918 to build it. The red brick building took five months to construct on the 800 block of East Center Street, a block west of the tent colony. The building included administrative offices, a dining room, nurse's quarters, and a kitchen. Army-style barracks were built behind the red brick sanitarium building for patients, which provided fresh air; they were a little better than the tents.

Voters approved another tax in 1938 to continue to support and expand the facility. A new, larger yellow brick building was added in 1939. It included rooms for patients. The wooden barracks were torn down. Dr. Pettit's privately owned sanitarium closed.

The entire world of T.B. treatment changed in 1944 when streptomycin was discovered; medicine now could fight tuberculosis. Open-air treatment was done, and sanitariums began to close. By the late 1950s, most were gone.

Dr. Vernon Madsen was medical director of the LaSalle County Tuberculosis Sanitarium from 1947 to 1969. Dr. Madsen contracted tuberculosis when he was serving his internship in an Iowa hospital. He was confined to a sanitarium for fifteen months, where part of the treatment was deliberately collapsing a lung, which left him with a weakened shoulder for the rest of his life. He decided to specialize in thoracic medicine—namely, the lung and the chest.

He met his wife, Clara, when they both worked in the University of Iowa hospital cafeteria. They married in 1931. A daughter, Beverly, was born in 1939.

Some patients did not like their confinement. Joanne White was in charge of telling Dr. Madsen whenever a patient escaped. She said Dr. Madsen was an imposing figure, "like a General," and many young nurses were afraid of him.

Treatment with drugs meant that sanitariums for long stays were not necessary. As the number of patients dropped, the state approved a measure in 1957 that allowed local sanitariums to use part of their facilities as nursing homes. The nursing home section of the building became known as Highland Sanatorium and Convalescent Home of LaSalle County. T.B. patients were moved to the second floor.

The LaSalle County Tuberculosis Sanitarium closed in 1969. The county sold the building, and it became a nursing home called Heritage Manor Nursing and Convalescent Home. It later was renamed Ottawa Care Center.

After the county sanitarium closed in 1969, Dr. Madsen managed a sanitarium in Rockford. He retired in 1972 and died in 1985.

The former sanitarium building and the old county detention home behind it were torn down in 2000. On that block, Ottawa Pavilion nursing home was built in 2001. It faces on Glover Street rather than Center Street.

A tuberculosis clinic was built in the early 1970s on Etna Road, on Ottawa's north side. In 1986, LaSalle County voters approve a measure to divert funding from the T.B. clinic to a county health department.

The LaSalle County Tuberculosis Sanitarium, 800 Center Street, Ottawa.

The house for the nurses at 907 E. Center Street, in the 1920s and in 2020.

Ottawa Care Center, in the former county T.B. sanitarium building.

Ottawa Pavilion.

8

TENTS AND TREATMENTS ELSEWHERE

Sanitariums for tuberculosis patients became prolific by the dawn of the twentieth century. This chapter looks at a few of them in Illinois and around the nation. The pictures here are important because they show scenes similar to what was happening in Ottawa.

In Illinois, a listing from 1908 showed the Chicago Home for Incurables, which opened in 1890 for advanced T.B. cases; Cook County Hospital for Consumptives at Dunning (which was the county poor house), which opened in 1899; Cook County Reception Hospital, which opened in 1907 for tuberculosis patients to be processed and sent to Dunning; the Open-Air Sanatorium for Jewish Consumptives in Chicago; and St. Ann's Sanatorium, which opened in 1903, was the first institution in Chicago devoted exclusively to the treatment of pulmonary tuberculosis. Chicago Tuberculosis Sanatorium opened a facility on 17 acres in Winfield, DuPage County, in 1908. Cook County's tuberculosis sanitarium in Oak Forest opened in 1910, but it only took transferred cases from Dunning.

Chicago's Municipal Tuberculosis Sanitarium was one of the largest facilities in the United States. There were eleven buildings on 160 acres on Chicago's northwest side, built between 1911 and 1939.

It was known by 1900 that people infected with tuberculosis should be isolated, but many sufferers could not afford private sanitariums. Public hospital beds for consumptives were in short supply. According to Chicago's Department of Health, more than 3,600 people died from tuberculosis in Chicago in 1905. The city's Department of Health appealed to Mayor Edward Dunne, saying that tubercular patients in a city hospital would be "a menace to the health" of others in the community. The health department told the mayor, "Give the Department of Health a sanitarium for tuberculosis and an increased working force and it is a certainty that the infection of tuberculosis will be less frequently conveyed from the sick to the well."

Dr. Theodore B. Sachs, one of America's foremost authorities in the control and treatment of tuberculosis, made a city sanitarium his goal. He already had established one of the nation's finest T.B. facilities—Edward Sanitarium in suburban Naperville.

Dr. Sachs was a Jewish immigrant from Russia who started out as a lawyer but switched to medicine. At his own expense, he investigated tuberculosis in the crowded

These pictures are from the Chicago T.B. Sanitarium. Above, children play in the Preventorium in 1926. Below is the open-air men's porch in 1925.

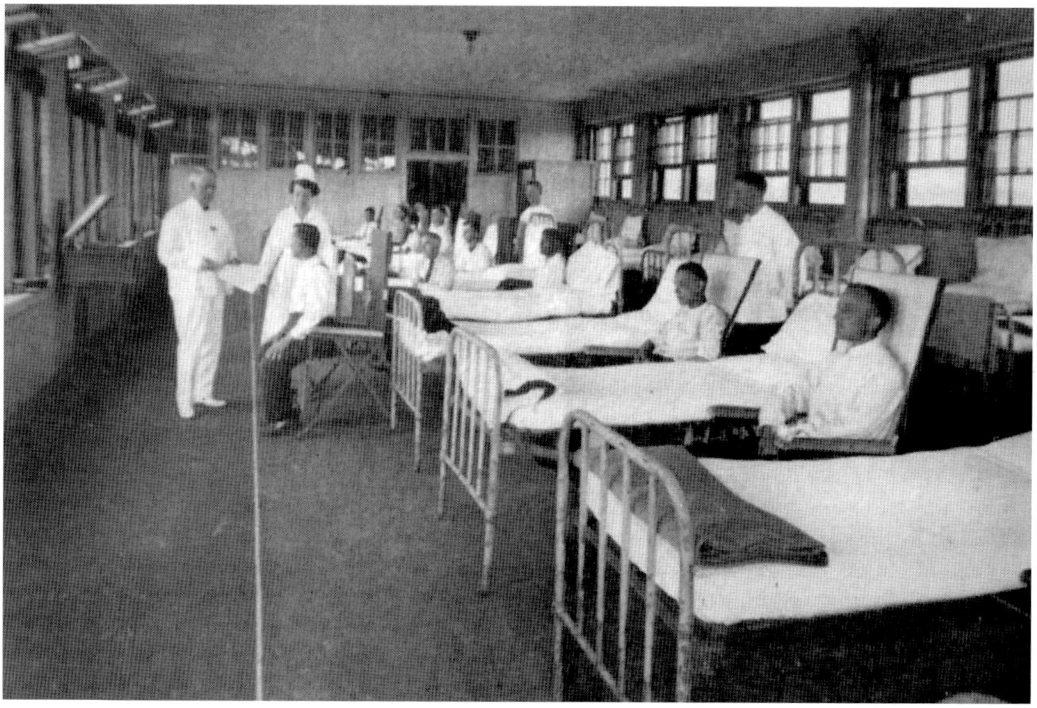

tenements of Chicago where the disease was rampant, particularly in the Jewish ghetto, which won him international acclaim.

Dr. Sachs was a founder and president of the Chicago Tuberculosis Institute in 1906. The group educated the public on tuberculosis, and it set up clinics, or dispensaries, in neighborhoods across the city. The goal also was to prove that Chicago's climate was as good as other settings. Dr. Sachs led experimental treatment at two new sanitariums just outside the city—at Gads Hill in Glencoe and Camp Norwood Tent Colony at the County Tuberculosis Hospital at Dunning. The doctor also worked at other area sanitariums, including Winfield Sanitarium, West Side Dispensary, and he restructured nursing care at the Cook County Sanitarium in Oak Forest.

Chicago Mayor William Busse appointed Dr. Sachs in 1909 to the board of the Municipal Tuberculosis Sanitarium, and he was named director in 1913.

State Senator Edward Glackin wrote a bill to establish state tuberculosis sanitariums. Since state funding was not likely, Glackin's bill made municipalities responsible for funding. It was signed into law in 1908 as the Illinois Public Tuberculosis Sanitarium Law, and it enabled municipalities to levy taxes to build and maintain their own sanitariums.

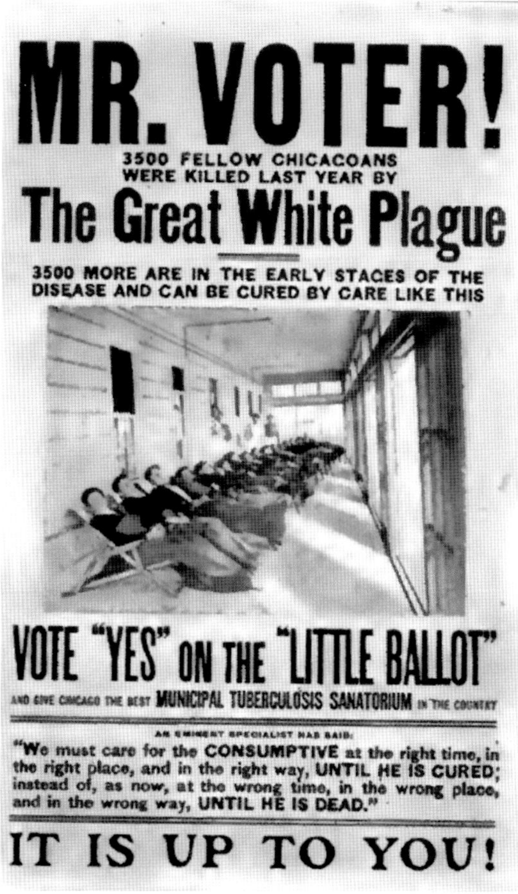

The city of Chicago held a referendum in 1909 to authorize the construction of a city sanitarium. Dr. Sachs and the Chicago Tuberculosis Institute campaigned for the measure. They were joined by health departments, the newspapers, labor unions, civic clubs, and others. The Chicago Tuberculosis Institute's plea was, "Tuberculosis is not so much a disease of paupers as it is a pauperizing disease; it is cheaper in the long run to cure the early stage consumptive and return him to his family than to care for him in a late stage in a pauper institution."

The voters approved the measure on April 5, 1909, which granted $1 million a year to fight tuberculosis and another $2.5 million to build the facility over four years. Mayor Busse appointed directors for the new Municipal Tuberculosis Sanitarium: Harlow Higginbotham, who had been president of the World's Columbian Exposition Corporation; Dr. William Evans, a previous commissioner of health; and Dr. Theodore Sachs as secretary. Drs. Sachs and Frank Wing were named to the Building Committee.

Dr. Philip Jacobs later wrote of Dr. Sachs:

Of all the many activities in which he engaged, none claimed so large a share of Dr. Sachs' personality and skill as the Chicago Municipal Tuberculosis Sanitarium. In a very real sense, the sanitarium was and is Dr. Sachs. It breathes his personality and his genius from almost every ward and brick. Into it he put his very body and soul.

A writer for the *Journal of the Outdoor Life* wrote in 1916, "The architect designed, but it was Dr. Sachs who breathed his genius through everything."

The Chicago Municipal Tuberculosis Sanitarium opened its doors on March 9, 1915, with a capacity of 650 beds and twenty-five buildings. In 1914, there had been over 3,900 deaths from tuberculosis in Chicago—16.3 for every 10,000 deaths. By 1924, the reported death rate from tuberculosis had fallen by nearly half, to 8.3 per 10,000 deaths. The institution closed in the 1970s after tuberculosis was brought under control in America.

EDWARD SANITARIUM

Edward-Elmhurst Health is one of the Chicago area's finest medical facilities today.

The original facility opened on January 15, 1907, in Naperville, Illinois, as Edward Sanitarium for tuberculosis patients. The capacity was for twenty-eight patients and the rate was $10 per week. It was founded by Dr. Sachs, who was the director and chief physician, and it was conducted by the Chicago Tuberculosis Institute. Financial help came from Eudora Hull Gaylord Spaulding. The institution was named for her first husband, Edward Gaylord.

Mrs. Spaulding added another $6,000 each year, which was enough to pay for ten patients who were too poor to pay for treatment. Mrs. Spaulding was impressed by the open-air treatment of T.B., which she had observed at the Glencoe Tent Colony, where she was a director of the Visiting Nurse's Association. Dr. Sachs also believed in the fresh air treatment and the fact that climate did not matter.

As the T.B. epidemic subsided due to antibiotic treatment and better public education on prevention, the facility in Naperville became a general hospital, renamed Edward Hospital, in 1955. It merged with Elmhurst Memorial Hospital in 2013.

Dr. Sachs was elected president of the National Tuberculosis Association in June 1915. Unfortunately, that was the same year William Hale "Big Bill" Thompson was elected mayor of Chicago. Thompson may be the most crooked politician ever in a city famous for crooked politicians. Along with Governor Len Small, Thompson was in league with Al Capone and other gangsters in a completely lawless era in Illinois.

Dr. John Dill Robertson, Thompson's city commissioner of public health, was forcing the T.B. sanitarium and other public health facilities to hire political appointees. Dr. Sachs, as head of the Chicago Municipal Tuberculosis Sanitarium, was one of the people targeted to be replaced. Thompson wanted his own political cronies in every position, even in medical positions for which they were not qualified. When he took office, Thompson would not reappoint Dr. Sachs, and he continually harangued him. The mayor finally was pressured to reappoint Dr. Sachs, but the harassment did not stop. Thompson said his reappointment of Dr. Sachs "was the worst I ever made," and he ordered his civil service commission to begin an inquiry.

Dr. Sachs became despondent and he resigned on March 20, 1916, saying that "the institution is being made a political football by the administration." Dr. Sachs wrote a letter to the board when he left, saying:

> I have refused to betray the community that has given me confidence. I have passed through ten months of continuous nightmare in trying to avert the politicalization of a great institution. But I find it impossible to continue. Single-handed at present, I cannot fight a big political machine.

He said the institution was "unsoiled with graft and politics," and he refused to turn it into "a city hall jobs factory." Dr. Sachs committed suicide by taking an overdose of morphine in his office in Edward Sanitarium on April 2, 1916. He was forty-seven. In his short suicide note, Dr. Sachs wrote:

> The Chicago Municipal Tuberculosis Sanitarium was built to the glory of Chicago. It was conceived in a boundless love of humanity and made possible by years of toil. No institution was ever planned more painstakingly or built more honestly. Every penny of the people's money is in the buildings, equipment and organization.
>
> The community should resist any attempt of unscrupulous contractors to appropriate money which belongs to the sick and the poor. Unscrupulous politicians should be thwarted. The institution should remain as it was built: unsoiled by graft and politics— the heritage of the people.

He concluded, "I am simply weary." His widow told the *Chicago Tribune*:

> The criticism of late affected him greatly. The doctor felt this great machine, City Hall, was crushing him but he didn't seem to realize, as I did, that the leading members of the profession and the public of Chicago were behind him.

The *Chicago Tribune,* along with a number of people in the medical community, blamed Big Bill Thompson for the doctor's death. DuPage County Coroner William Hopf said at

Dr. Theodore B. Sachs and his grave marker in Naperville.

the inquest, "His suicide was due solely to persecution. The result constitutes a political murder. That is a strong phrase but it is the way I feel."

Praise for Dr. Sachs came from every corner of Chicago, especially from the medical and the religious communities. Mayor Thompson ordered his aides to keep silent.

An article in the University of Illinois Alumni Quarterly said:

Tuberculosis he would have fought to the last. But that tuberculosis of municipal morals known as spoils politics, he could not understand. He felt the virus of City Hall politics tainting the great municipal tuberculosis sanitarium. That institution meant everything to him. All his hopes for Chicago's tubercular sufferers were there. All he had went into it—energy, ability, sympathy, ambition. Offered $10,000 a year to take a private position, an honorable one, too, he shook his head and went back to the work he loved and for which he received no pay. As money goes, he died a poor man.

He abandoned all generalities and specialized in tuberculosis treatment. He neglected a private practice that would have made him rich in money. He devoted much energy to the Edward Sanitarium for tuberculosis at Naperville, a small private infirmary of which he was director, and he went back there to die. "It is the only place where I have known peace in life" were his mournful words.

Dr. Theodore Bernard Sachs was buried on the grounds of Edward Sanitarium.

Dr. Sachs met Louise Wilson in 1900 when he was a doctor and she was a nurse at Michael Reese Hospital in Chicago. She developed tuberculosis, and he took her to a sanitarium in Denver and treated her for three years. They married when they returned to Chicago.

Mrs. Sachs took a job at Edward Sanitarium after her husband's death. She died in Chicago in 1967 at the age of ninety-two.

A *Chicago Tribune* article in 1917 on the tenth anniversary of Edward Sanitarium showed that 66 percent of former patients were back at work. Several hundred people who had been cured of tuberculosis attended a homecoming celebration on June 8, 1952.

Chicago has been notorious for its corruption ever since the city was established. The corruption has extended to every facet in Chicago, even to the charities and the medical institutions devoted to the public good. Dr. Sachs was one of its victims, and it was a burden he found too hard to bear. Dr. Frank Billings had blasted Chicago politics at the Illinois State Medical Society's annual meeting in Quincy in May 1909. The *Chicago Tribune* reported:

> Dr. Billings objected to political methods and ideas in the care and control of state institutions and urged the 8,000 physicians in Illinois to work for both scientific and practical management of the public institutions along lines as exemplified in New York. This work should begin in the selection of members of the general assembly, he said.

The subject of Chicago's perpetual pay-to-play politics again was raised at a meeting of the Chicago Equal Suffrage Association in February 1918. "There are certain positions in Chicago which seem to be sacred to temporary employees," Mary Rozet Smith told the group. She said the Chicago Municipal Tuberculosis Sanitarium had 331 temporary workers of its 550 employees. Mrs. James Morrison said this was a violation of the civil service law. The *Chicago Tribune* reported, "A little woman with gray feathers in her hat remarked naively, 'I thought it was done to provide more plums.' Amelia Sears noted, "It was different in the time of Dr. Sachs."

Other examples are far too numerous to detail here. It still happens on a daily basis.

North Carolina built a number of sanitariums. Highlands Sanitarium, built in 1908, used the fresh air method of treatment. It was locally known as "the San" or "Bug Hill." Outside the main building were sixty tents and wooden frames with canvas tops and sides. However, in February 1918, a workman thawing frozen pipes in the main building accidentally burned down the building. That was the end of the colony. The property later became a trailer park. One of the tents was still being used as a utility shed. The Highlands Historical Society restored it in 2006, and it sits on Bug Hill, which is now a public park.

Arizona became a haven for people with tuberculosis and other lung ailments because of its hot, dry climate. Dr. Mark Rodgers wrote in 1896 in a medical journal, "The climate of Arizona is peculiarly adapted to the requirements of people suffering from pulmonary disease."

Tuberculosis probably brought more people to Arizona than mining, ranching, farming, or commerce, wrote Jo Baeza in *The Arizona Independent* in 2011. "Tuberculosis was endemic in the 1880s and 1890s, the leading cause of death across the temperate zones of the world. Health was big business and Arizona promoted its climate zealously," she wrote.

Edward Sanitarium, Naperville.

An aerial view of Edward Sanitarium.

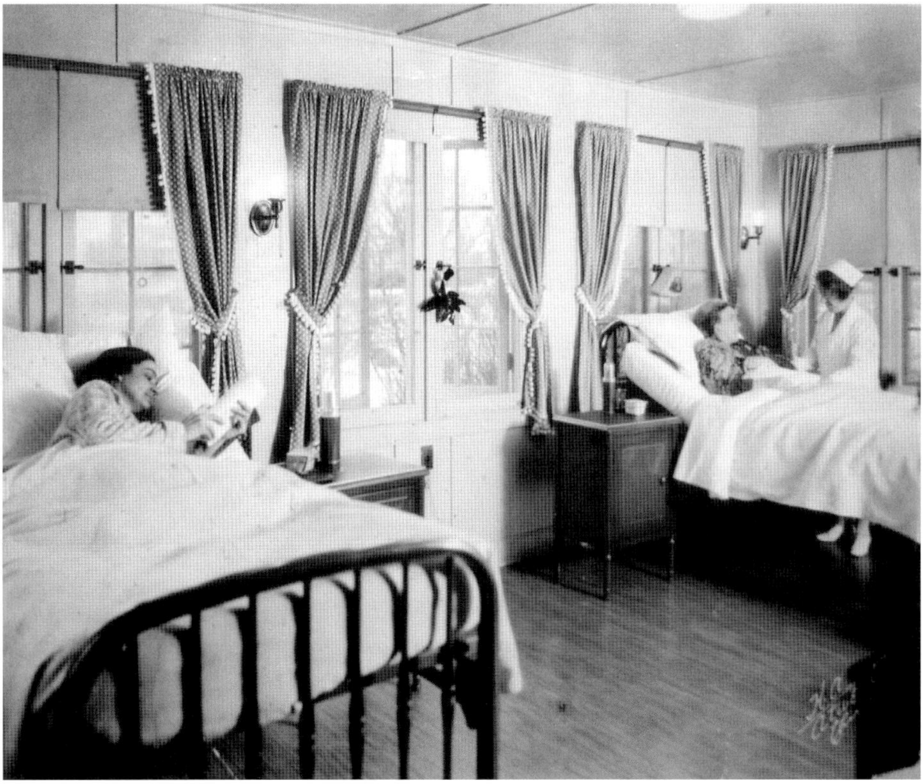

Tuberculosis patients in Edward Sanitarium.

Tubercular isolation huts under construction in Alabama in the 1930s.

Approximately 70 percent of the Arizona territory was infected with tuberculosis in 1900, according to the Arizona Medical Association. Frances Quebbeman, in her article "Medicine in Territorial Arizona" for the Arizona Historical Foundation in 1966, said only 10 percent of these cases progressed to an active state; many patients continued working and some never knew they had it but continued to spread it. For the poor, there was Tent City, also called Lunger Hill, in the desert north of Tucson in the 1920s.

Famous gambler and gunfighter John Henry "Doc" Holliday was working as a dentist in Atlanta, Georgia, when he was diagnosed with tuberculosis. His mother had died of the disease. He went west for a cure in the 1870s, first to Dodge City, Kansas, and then to Tombstone, Arizona. He joined his friend, Wyatt Earp, in the infamous gunfight at the OK Corral, and the rest is Western history and myth. Doc Holliday died in 1887 at the age of thirty-six in a tuberculosis sanatorium in Glenwood Springs, Colorado.

Colorado also promoted its climate as ideal for curing tuberculosis. In fact, a good number of the people who moved to Colorado and helped build the state were people who settled there seeking a cure. The promotion started in the 1870s. The Colorado Health Resort claimed in 1883:

No other locality in the known world has a climate equal to that of Colorado Springs. Our air is pure and bracing, and is entirely free from all malarial influences. Persons suffering from chronic consumption are likely to live longer and more comfortably by residing in Colorado.

Outdoor rest, sunshine, and fresh air in an Arizona facility.

Health resorts were built, and parks and roads and other development boomed as people flocked into the state. The 1883 brochure said, "There is absolutely no need of dying of consumption if one will try the climatic prevention." The Colorado sanitariums followed the same regimen as Dr. Pettit—fresh air, rest, and a nutritious (and fattening) diet. The promotions worked. Leah Davis Witherow, the curator at the Colorado Springs Pioneer Museum, wrote, "By 1900, approximately 20,000 health-seekers emigrated to the southwest each year, with one-third of Colorado residents coming to the state in search of a cure for themselves or a close family member."

Matt Mayberry, director of the Colorado Springs Pioneers Museum, wrote, "Tuberculosis was our first major industry in Colorado Springs. We were really just a resort town but tuberculosis became the major driving force of our economy from about the 1880s until after World War II."

In the 1880s and 1890s, one-third of the people in Colorado Springs had tuberculosis. It became a problem in Denver because many who sought a cure by coming there were poor and homeless. Some were taken to jail. Some were so ill when they arrived that they died soon. Boarding houses on North Nevada Avenue were called "Lunger's Row."

By 1917, Colorado had more than a dozen sanatoriums, and each one had a number of huts. The Modern Woodmen of America Sanatorium in Colorado had 245 individual wooden huts at one time. The M.W.A. Sanatorium closed in 1947, and more than 200 tent cottages were sold. Many can still be seen in backyards as tool sheds and more.

The National Jewish Hospital for Consumptives opened in Denver, Colorado, in 1899. It was founded by a group of immigrant Eastern European Jewish men who were survivors of T.B. The sanatorium treated tuberculosis patients in all stages of the disease, free of charge. Its motto was, "None may enter who can pay, none can pay who enter." While it was originally built by the Jewish community, anyone who needed help was admitted.

Initially housed in wooden "tent" cottages, the patients were given the benefits of fresh air and wholesome kosher food. It was one of the leading tuberculosis sanitariums in the country at the turn of the twentieth century, in a city already known as a leading center for T.B. sanitariums.

The facility used heliotherapy, which was sunlight therapy and open-air therapy. Their studies showed T.B. patients recovered faster in the winter than in summer months:

Dr. Otto Walther says he not infrequently had an inch of snow on his blankets. At some localities, tents are the only shelter, but whether the patients are housed in cottages or in tents, the free access of air must be absolute and uninterrupted. Drafts are not feared. At night the windows, which should constitute at least one side of the room or ward, are kept open; in some places the sashes are removed altogether; the sides of the tents are rolled up, except in the severest storms. In very cold weather, the head and hands may be protected with woolen cap and gloves, and at all times the patients are well provided with blankets. In summer, the beds or reclining chairs are moved into the open or into covered porches during the day; for the winter, most places are provided with glass porches or sun parlors where the patients spend their days in bed or reclining in [a] steamer chair.

Cough and night-sweats disappear in a short time, and, as a logical consequence, the medicinal treatment is reduced to a minimum. Antipyretic drugs are never used and expectorants are rarely required. Suralimentation [the belief that T.B. could be cured by forcing patients to eat] is practiced in many places, especially in the German resorts, where it is pushed to an almost incredible degree. Even bed patients with considerable pyrexia are placed on a full diet of meat and vegetables.

Previous page and above: Two views of the M.W.A. sanitarium and tent colony in Colorado.

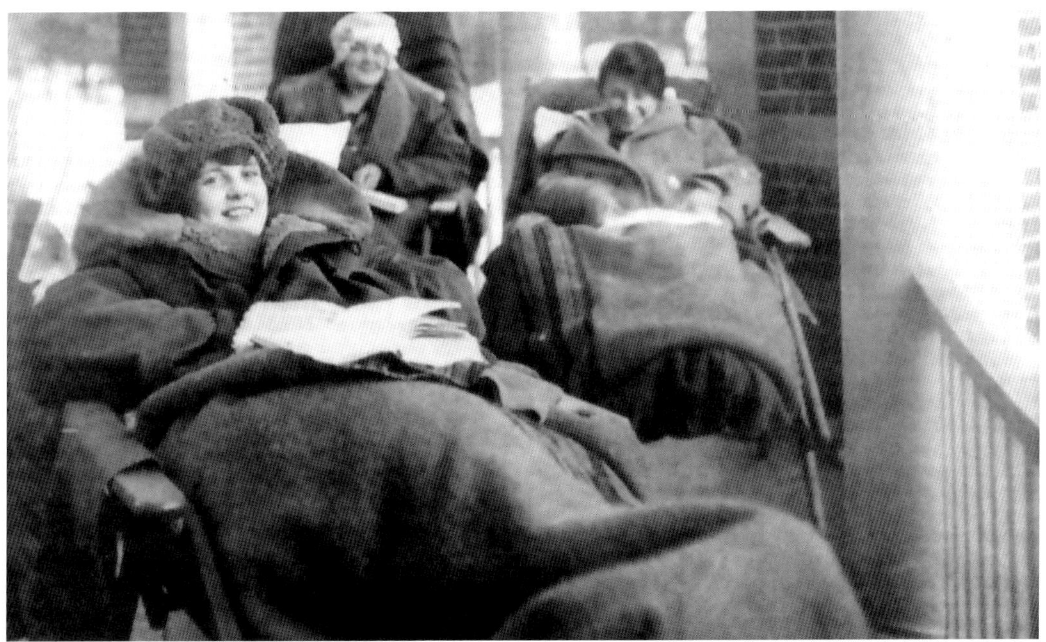

This is how patients in Ottawa and other tent colonies would look—resting and wrapped in blankets to stay warm while open to breathing fresh air at all times. These patients are at the sanitarium in Raybrook, New York.

T.B. patients get sunlight therapy at the Jewish Consumptives Relief Society sanitorium in Denver, Colorado, in the 1930s.

In 1954, The National Jewish Hospital for Consumptives changed its mission to cancer research and became the American Medical Center. Today, it is known as the A.M.C. Cancer Research Center.

The New York State Hospital for Incipient Pulmonary Tuberculosis opened in 1904 at Raybrook, in the Adirondacks, near the experts at Saranac Lake.

OPEN-AIR SCHOOLS

Treatment went beyond tents and sanitariums. There were quite a number of schools that held open-air classes, even in the winter.

School opened on August 3, 1909, at Seventy-Sixth Street and Vincennes Avenue in Chicago, in what is now the West Chatham neighborhood. Twenty-six children from the stockyards district who had mild cases of tuberculosis began an experiment of outdoor classes. The day began with washing faces and hands, followed by a nutritious breakfast. There also was exercise, sunbaths, and an hour of sleep. There were plenty of meals to build up a body that otherwise might waste away.

Outdoor schoolrooms—above left, in New York in 1900, and above right, in 1910 in Providence, Rhode Island.

By 1911, there were three outdoor schools in Chicago for sick children. A *Chicago Tribune* article on August 13, 1911, began:

In three schools of Chicago, oatmeal and corn are substituted for arithmetic and spelling. Gravies, carefully prepared with regard to the laws of dietetics, and meats with a high food value are substituted for history and writing. Rest and sleep replace tiresome lessons in reading and geography.

Thirty-three children in one school gained 57 pounds in one week.

A tuberculosis epidemic in 1908 led to an experiment in Providence, Rhode Island. Children at one school sat in their classroom, bundled against the freezing weather, in an open-air school, patterned after tent colonies like the one in Ottawa. This was the first school in the country to fight T.B. in this manner.

There was a high rate of T.B. at that time in Rhode Island. Drs. Mary Packard and Ellen Stone made the suggestion for the open-air school. The students were from Dr. Packard's summer camp for tubercular children, and she did not want them infecting other students in school that fall. Ginia Bellafante's story in the *New York Times* in 2020 said the children were in wearable blankets with heated soapstones at their feet. "The experiment was a success by nearly every measure—none of the children got sick." Within two years, she wrote, there were sixty-five open-air schools around the country based on the Providence model.

Students sat in direct sunlight. They had breathing exercises to build up their lungs. The area was heated, even though fresh air blew in. The experiment was deemed a success. The children gained weight and their health improved. A second open-air school was opened in 1913. Dustin Waters, writing in the *Washington Post* in 2020, said that ultimately, eleven such schools began and continued for another thirty-one years. "By that time, the concept of an open-air classroom had spread to more than 150 American cities, aiding both the minds and bodies of thousands of students." Open-air schools in Providence ended in 1957, largely because the disease was being treated with antibiotics instead.

Chicago schools began open-air education to combat T.B. when the disease was reaching epidemic proportions in the early 1900s. School officials took the example of

Students taking sun therapy. (*American Lung Association*)

places like the Ottawa Tent Colony and brought the classrooms outdoors. The *Tribune* reported in 1909:

> The mostly outdoor school, led by teachers and nurses, emphasized physical activities 'in an attempt (for the students) to acquire strength to combat the incipient stages of the white plague. First of all, the face and hands of each one were washed, and then the children sat down to breakfast. Then the children washed the dishes, took their first lessons in camp duties, and received light instruction in the school building, the tent in which the classes are to be held not having been delivered. There were frequent intermissions for play, story telling, and gymnastics.
>
> After luncheon, there was an hour of rest, many of the children taking sun baths in large canvas chairs. After more gymnastics there was another hour for sleep, then play, more rest, and supper. The dishes were washed again, each had a shower bath, and, provided with car fare, the pupils were sent to their homes, to return this morning.
>
> The plans of the promoters of this school contemplate the opening of winter schools with all windows and doors removed. Tuberculous children will sit wrapped in blankets, furs and mittens while they study.

Apparently, the open-air school made it through its first winter, even though the open-air classrooms in winter were not approved by the Chicago Board of Education.

Principal William Watt told a Board of Education meeting in October 1910, "We disobeyed the board's rules, but we didn't say anything about it until the experiment proved successful." Watt added, "We let the children wear overcoats and overshoes and they were as warm as they wanted to be, with plenty of exercise and something to keep them interested." Watt said a large number of people wanted more open-air classrooms. Three other open-air schools were operating by 1911.

A *Tribune* editorial endorsed the methods:

> Any deficiency in learning is so amply compensated for by an increased rosiness of cheek and plumpness of arm that no distress is expressed by the strictest of formal educators. They are schools in which efficiency is gauged by gain in appetite, not in mental achievement, but this does not imply that they exactly neglect the cultivation of the mind.

Sherman C. Kingsley, superintendent of the United Charities of Chicago, was an advocate of the open-air method:

> The universal testimony of these open air schools is that the children gain in weight, that temperatures are reduced, that listless children become alert and attentive and that there is a marked change in the mental grasp of the children. In the Elizabeth McCormick Open Air School there was an average gain of over four pounds.

Writing about it in the *Chicago Tribune* in 2020, Elise De Los Santos said that the open-air idea was coming back:

> An outdoor pre-school ran for about five years in a wooded area of the Chicago Park District's North Park Village campus. It was modeled on the idea that unstructured play

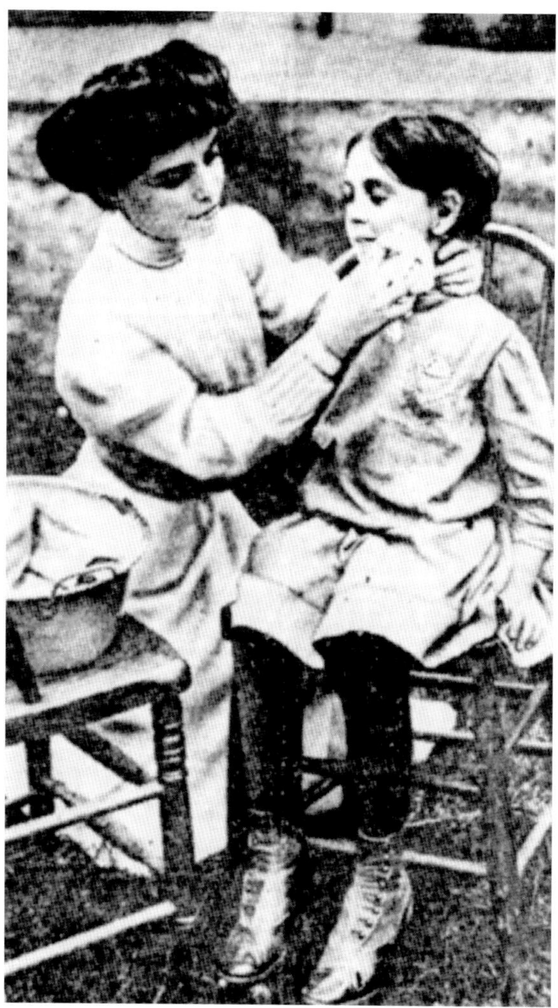

Nurse Harriette Leland washing
a Chicago pupil in 1909.

in nature teaches essential skills and also that outdoor education combats modern ills in children like obesity and depression. It was shut down last August after complaints from neighbors. Coincidentally, North Park Village used to be the Chicago Municipal Tuberculosis Sanitarium complex, once one of the largest facilities in the U.S. dedicated to treating those with the disease.

Amelia Nierenberg, writing in *The New York Times* in 2020, said similar outdoor learning was being done in several parts of the country because of the COVID-19 pandemic. The idea came from what had been done more than a century earlier in the tuberculosis epidemic and the Spanish Flu epidemic. These included public schools in Falmouth, Massachusetts; Essex Street Academy in New York City; Lakeside School District in Hot Springs, Arkansas; and Prairie Hill Waldorf School in Pewaukee, Wisconsin.

Outdoor public school in Manhattan, New York, in 1911.

A scene at the Penrose Tent Colony in Colorado Springs, Colorado, in 1913.

9

THE PANDEMICS AND THE MEN WHO FOUGHT FOR THE CURES

There have been a number of deadly epidemics and pandemics throughout history, including influenza, pneumonia, malaria, cholera, dysentery, scarlet fever, yellow fever, smallpox, measles, and more. These epidemics would kill thousands of people or more and then go away for a number of years until immunity faded, and then they would return to kill thousands more. Tuberculosis did not act that way. It was always there, killing without ceasing. This chapter briefly looks at a few of these diseases and the men who fought tuberculosis.

The Bubonic Plague, or the Black Death, took an estimated 75 to 200 million lives in Europe, Eurasia, and North Africa in just seven years, beginning in 1346. It is estimated that perhaps half the population of Europe died. The disease was caused by a bacteria brought from Central Asia, carried by fleas on rats. A lack of hygiene caused it to spread among the human population rapidly, and death was swift. It was so devastating that even today the name "The Plague" brings this pandemic to mind. There were minor outbreaks of this strain before and after this disaster. This was only a couple of decades after the Great Famine, which killed between 30 and 60 percent of Europe's population.

The Antonine Plague, which lasted from 165 to 180 A.D., was brought back to the Roman Empire by troops returning from battles in the Near East. The plague may have been smallpox and measles or may have been a strain of the Bubonic Plague, and it killed approximately 5 million people, which was about one-third of the affected population. It devastated the Roman army, the population, the economy, and every other aspect of life. The Plague of Justinian, in only eight years from 541 to 549 A.D., was caused by the same bacterium that caused the Black Death and is believed to have killed a quarter of the people in the eastern Mediterranean area. As many as 5,000 people died every day in Constantinople at its peak. The Russian flu pandemic (1889–1894) is believed to have killed 1 million people.

The Spanish Influenza pandemic (a strain of the H1N1 virus) swept across the world in 1918 and 1919, killing approximately 50 million people, including 675,000 in the United States. More than a third of the world's population, approximately half a billion people, came down with the flu. There was no testing, vaccine, or ventilators at the time, and an insufficient treatment was all that could be done.

More recent pandemics include a polio epidemic in the first half of the twentieth century that killed thousands of people and crippled tens of thousands before a vaccine was developed by Dr. Jonas Salk in 1955; Dr. Albert Sabin developed an oral polio vaccine in 1961. The Asian Flu of 1957 killed 1 million people worldwide and 80,000 in the United States. The Hong Kong Flu of 1968 is estimated to have killed between 1 and 4 million people worldwide and 100,000 in the United States. In the H1N1 virus pandemic of 2009 to 2010, the Centers for Disease Control (C.D.C.) gave a wide estimate of the worldwide death toll between 151,000 and 575,000, including 12,469 in the United States. The World Health Organization (W.H.O.) said there were 207 million cases of malaria worldwide with 627,000 deaths in 2012. It also said there are 3–5 million cases of cholera a year. The death toll from COVID-19 in 2020 and 2021 is great, but exact numbers are unknown because reported figures are unreliable.

The American Review of Tuberculosis reported there were 153,000 T.B. deaths in America in 1900. That number fell to 100,000 in 1925 (even though the population had increased by 40 million).

The Illinois Department of Public Health reported 372 cases of tuberculosis in the state in 2010 (California had the highest that year with 2,329 cases, followed by Texas with 1,385, New York with 955, Florida with 835, Georgia with 411, and New Jersey with 405).

There has been a resurgence of tuberculosis in recent decades because of drug-resistant strains of the bacterium and the emergence of H.I.V. This led the World Health Organization to issue a declaration of a global health emergency in 1993. It is estimated that 500,000 new cases of drug-resistant tuberculosis cases occur annually worldwide.

W.H.O. said there were 9 million cases of tuberculosis in 2013 and 1.5 million deaths. Tuberculosis is the leading killer of H.I.V.-positive people, causing one-fourth of H.I.V.-related deaths.

THE SCIENTISTS

A number of men in the 1600s and 1700s, such as Franciscus Sylvius, Benjamin Marten, Leopold Auenbrugger, Robert Whytt, Percival Pott, and Benjamin Rush, made advancements in the study of T.B.

René Laennec invented the stethoscope in the early 1800s, which allowed a better diagnosis of what was happening in the lungs. He died from T.B. at the age of forty-five, after contracting it while studying contagious patients. Jean Antoine Villemin demonstrated in 1865 that T.B. was an infectious disease.

The first idea for a sanatorium was made by George Bodington in 1840, who proposed a dietary, rest, and medical program. Dr. Hermann Brehmer contracted tuberculosis in the 1840s. He opened a hydrotherapy institution in Germany in 1854, the first sanatorium. It was so successful that similar sanatoriums opened across Europe and the United States. By the late 1870s, the sanatorium movement began to spread throughout Europe.

"Sanitarium" and "sanatorium" are both used, but strictly speaking, a sanatorium is a hospital-like facility providing care for tuberculosis and other chronic diseases, whereas a sanitarium is often a spa-type facility where someone might go to recover their health.

Dr. John Croghan brought fifteen tuberculosis patients into Mammoth Cave in 1842 on the theory that they could be cured with a constant temperature and purity of the

From left to right: H. Robert Koch; Wilhelm Roentgen; and Edward Trudeau.

On the bluff outside the Club House, with the Illinois River and the city of Ottawa in the distance.

cave air. Patients were kept in stone huts. However, two patients died and the others left and soon died. Dr. Croghan died of tuberculosis in 1849.

The name tuberculosis was first used by Johann Lukas Schonle in 1839. It is from the Latin "tuberculum," meaning a small swelling or bump. The word became into use in 1882 after Dr. Koch's discovery of the tubercle bacillus.

German physician and microbiologist Dr. Hermann Heinrich Robert Koch was the first to identify the bacteria that caused diseases such as cholera, anthrax, and tuberculosis. His research, published in 1882, was instrumental in developing "germ theory" and it caused changes in sanitation and helped create guidelines for the control of diseases. During this time, T.B. killed one out of every seven people living in the United States and Europe. Dr. Koch announced his results at the Physiological Society of Berlin on March 24, 1882, and that is why that day has been known as World Tuberculosis Day since then. Dr. Koch received the Nobel Prize for Medicine in 1905 for his work in tuberculosis.

Dr. Edward Livingston Trudeau founded the National Association for the Study and Prevention of Tuberculosis in 1904. In addition to studying the disease, the organization was dedicated to educating the public about prevention and to research in finding a cure. Today, the group is named the American Lung Association.

Dr. Trudeau contracted T.B. in 1873 at the age of twenty-six and was given six months to live. He built a shack in the Adirondacks in 1880 where he and two other sufferers sat in below-zero weather wrapped in blankets. It was successful, and he lived to the age of sixty-seven.

Dr. Trudeau established the Adirondack Cottage Sanitarium for T.B. patients at Saranac Lake, New York, in 1884. It was the first American tuberculosis sanitarium and a research facility for T.B. patients. It became the model for the development of hundreds of similar private sanitariums across the country. At a time when health spas catered to the wealthy, Dr. Trudeau's open-air cottage sanitarium was the first facility for the poor. Dr. Trudeau's brother and his son and daughter died of T.B.

Dr. Trudeau wrote in 1887 that a healthy outdoor lifestyle with proper nourishment could almost eliminate the disease. He based his therapy on Dr. Brehmer's "rest cure" in cold, clear mountain air. He continued Dr. Koch's experiments and built the best-equipped medical research laboratory in America. Today, the Adirondack Cottage Sanitarium continues as the Trudeau Institute.

In 1895, Wilhelm Konrad von Roentgen discovered the x-ray, which allowed doctors to see T.B. in its early stages instead of waiting for a diagnosis when it is too late. For this, he won the 1901 Nobel Prize in Physics in 1901.

In 1944, Selman Waksman developed streptomycin, the first effective antibiotic against tuberculosis. After tests, it was given to a patient in 1949, and the patient was cured. However, the bacteria developed resistance to the drug in many people. The development of isoniazid in 1952 and rifampin in the 1970s hastened recovery times and reduced the number of tuberculosis cases.

Antibiotics and education and prevention caused a steep decline in cases, and sanitariums began closing in the 1950s. A.G. Holley Hospital in Lantana, Florida, was the last tuberculosis sanitarium in the United States, closing in 2012.

Bundled up and resting at Trudeau's Saranac Lake sanitarium.

Above left: The Christmas seals program started in 1907 as a means to raise money to fight tuberculosis.

Above right: The 1936 poster warned of spreading tuberculosis germs.

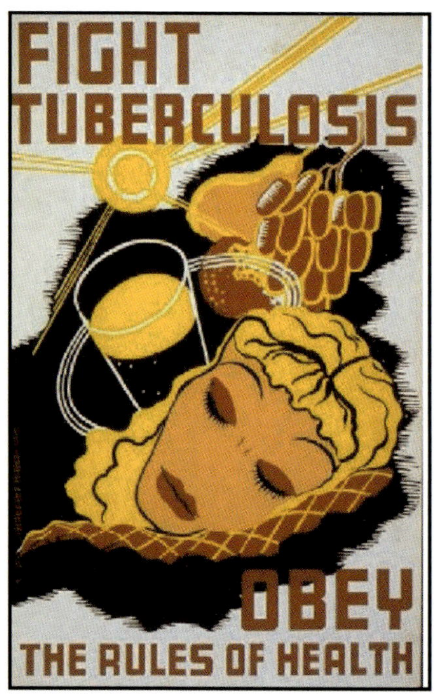

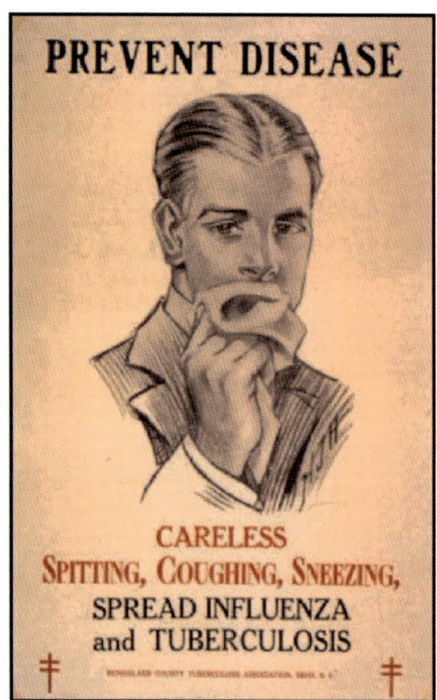

Posters from before World War I through the 1940s promoted T.B. prevention.

THE NEXT TO GO
FIGHT TUBERCULOSIS
Red Cross Christmas Seal Campaign

A WAR ON CONSUMPTION
Printed and distributed by the
Metropolitan Life Insurance Company

TUBERCULOSIS AND CHILDHOOD
The State must protect its children.
Almost 50,000 American children die
yearly of tuberculosis. The anti-tuberculosis campaign must concentrate on
the child.
Almost all children have dormant tuberculosis germs in their bodies. To keep
these dormant germs from developing active
disease, train children in the proper use
of
FRESH AIR - SUNSHINE
GOOD FOOD - REST

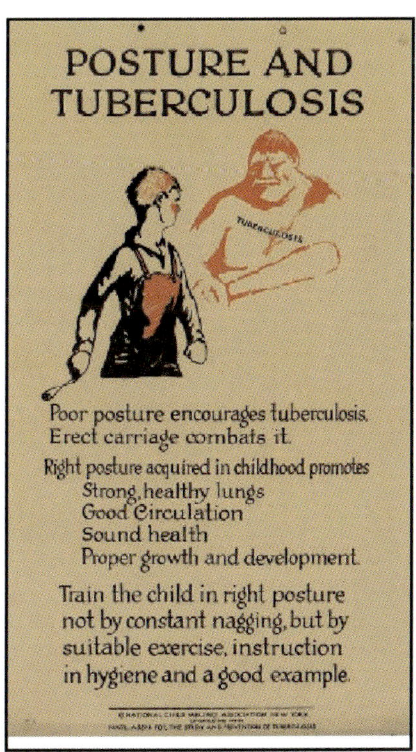
POSTURE AND TUBERCULOSIS
Poor posture encourages tuberculosis.
Erect carriage combats it.
Right posture acquired in childhood promotes
Strong, healthy lungs
Good Circulation
Sound health
Proper growth and development.
Train the child in right posture
not by constant nagging, but by
suitable exercise, instruction
in hygiene and a good example.

10

The U.A.W. and the Tent Colony Property Today

The 14-acre tent colony property on East Center Street in Ottawa was sold in 1949 to the United Auto Workers union as a regional educational center. The purchase was arranged by Pat Greathouse, the U.A.W.'s Region 4 director. The purchase price was $30,000. All the buildings eventually were razed except the historic Club House.

This was an era when African-Americans were not welcome in many places —not in segregated Chicago and not in the South. Even at union conferences, African-Americans could not go to certain restaurants and hotels when union members met. The U.A.W. opposed segregation and began looking for a centrally located property where they could welcome everyone regardless of race. The Ottawa property seemed ideal. It was, according to a 2017 story in the *Ottawa Times*, "Greathouse's goal to break down segregation by bringing members together to learn, eat and lodge free of the discrimination they faced in hotels and restaurants. Because of its proximity to the South, the center was a safe haven for members during the violent and dangerous Civil Rights era."

The center helped create unity among workers in the far-flung areas of Region 4. It also brought together people of all races.

The facility was dedicated on Labor Day in 1949, with Illinois Senator Paul H. Douglas on hand. It was named the Ottawa Union Center. It was renamed the John F. Kennedy Union Center after President Kennedy's assassination in 1963. It became the Pat Greathouse Education Center in 1994.

Mr. Greathouse was born in 1915 in West Salem, Illinois. In 1956, he married Marguerite McCandless in Washington, D.C. They moved to Detroit, and he served as vice president of the International Union, United Auto Workers, from 1956 to 1980. He also was organizational director of the union and was involved in contract negotiations and grievance handling. He traveled extensively in the United States and around the world, working with labor leaders and government officials in many countries.

After becoming a vice president, he continued to be active with the U.A.W. Region 4 Education Center in Ottawa. Mr. Greathouse died in 2005.

A hotel was built in 1957 by Glen Vissering of Streator. Guests stay in the hotel while attending events, conferences, and training at the education center. Meals are served in the historic dining room in the Club House. The center serves approximately 2,500

Pat Greathouse.

U.A.W. members a year. Besides activities for the U.A.W., events include police and fire union meetings, family picnics for the members, postal worker union meetings, retirees, veterans, civil and human rights conferences, and training.

The hotel was built where the tents and huts had stood. The huts were sold for sheds, fishing shacks, garden sheds, and more. One of the huts, which was being used as a garden shed in Utica, was brought back to Ottawa in 2019 and was restored by union workers.

In addition to the Club House and the hotel (with 1,000 feet of river frontage), there is an education building, an in-ground swimming pool, a pool house, and a large garage.

Walter Reuther was so impressed with the Ottawa center that it inspired him to create the fabulous education center at Black Lake, Michigan. Mr. Reuther headed the U.A.W. from 1946 until his death in a plane crash in 1970. He was one of the most powerful labor union leaders in American history, taking a fledgling union and building it into a powerhouse.

Mr. Reuther's support helped make possible the Peace Corps, the Civil Rights Act of 1964, the Voting Rights Act of 1965, Medicare, and Medicaid. Mr. Reuther was a friend of Dr. Martin Luther King, Jr. Reuther marched with Dr. King in Selma and other places. When Dr. King was arrested in Alabama, where he wrote his famous "Letter from Birmingham Jail," it was Mr. Reuther who put up the bail money for Dr. King. Reuther also helped organize and finance the 1963 March on Washington where Dr. King gave his historic "I Have a Dream" speech.

The building to the west, formerly used by Ottawa General Hospital and the Arthritis Clinic, was bought by the U.A.W. in 1985 from the city. It was going to be used for classrooms but it was full of asbestos and there were other problems. It was torn down a short time later and the land remained vacant until 2020 when construction of Region 4 Headquarters began. It will be completed in 2021.

147

The entrance to the U.A.W. Union Center in the 1950s.

The motel and swimming pool at the U.A.W. Center in Ottawa.

Walter Myers, who managed the U.A.W. property from 1955 to 1974.

The first U.A.W. hotel, built in 1957.

Dr. Pettit's house a block away at 823 E. Center Street was purchased in 2016. It had been a private home for years. The U.A.W. decided it did not need it, and it was sold in 2020 as a private residence.

A tornado on February 28, 2017, devastated parts of Ottawa and Naplate. The U.A.W. hotel was one of the buildings destroyed. The wreckage was removed and a brand new hotel was built on the site. It increased the number of rooms from thirty to fifty. Insurance paid for what was lost, and the balance was provided by contributions from union members. The new hotel opened in May 2018. Brad Dutcher, Region 4 assistant director, said everything was locally sourced, from plumbing fixtures to the furniture purchased from an Ottawa store. Mr. Dutcher said:

> As soon as they found out about the devastation, Region 4 members came to the rescue of the center and nearby communities. U.A.W. members have a history of cleaning, landscaping and repairing the center. Members felt the destruction on a personal level. They picked up debris, cut trees, and salvaged what they could. We had so many volunteers they started helping communities clean up, too.

"When we were devastated by the tornado, we were there for each other," Region 4 Director Ron McInroy told *The Times* of Ottawa. "We cleaned ourselves off and began rebuilding. The U.A.W. has always preached the importance of community involvement. A time like this is why. We need one another."

"The tornado may have caused destruction, but it's brought our members and their community closer. I'm thankful we still have a future here despite our setbacks. Our union is strong and our spirits are high," said Mr. McInroy.

Today, the 1905 Club House is restored to a near-perfect condition. The U.A.W. has respected the history and tradition of the tent colony by keeping the Club House as original as possible. Pictures on the walls show the tent colony in the earlier era. The dining room looks as it did more than a century earlier. Large windows give diners a look at the beautiful Illinois River below the bluff, with views of Scherer and Bulls islands and a large forested area with the city of Ottawa in the distance.

DISTINGUISHED GUESTS

The U.A.W. center has hosted some important people over the years. John F. Kennedy campaigned there in 1960. Vice President Al Gore visited. Barack Obama spoke there during his campaign for president.

Dr. Martin Luther King, Jr., and his wife, Coretta Scott King, stayed there in the 1960s, using the center to plan civil rights strategy with their aides. It was one of the only places in the region at the time where blacks and whites could stay together in peace.

Rich Myers remembers Dr. King from his stay there. Rich grew up on the tent colony property. His father, Walter, was manager of the U.A.W. facility from 1955 to 1974, and they lived in the Club House. Rich Myers spent his career in education in LaSalle County, becoming regional superintendent of schools. He was fortunate enough to meet and talk with Dr. Martin Luther King, Jr., when the civil rights leader stayed there. Rich

The new U.A.W. hotel, which opened in 2018.

The new U.A.W. Region 4 headquarters, which will be completed in 2021.

remembers that he was home for the summer from Southern Illinois University at the time, so it would have been the summer of 1965 or 1966. Here is part of an interview he gave to reporter Charles Stanley for *The Times*:

> Dr. King was in Ottawa to talk because a lot of the members of the local unions were supporting his freedom marches. So the Kings stayed with us and had dinner with us. He was very congenial and very personable. He was very much a person who enjoyed talking about his life and his work. He just seemed like a very positive man. Of course, at that time, his claim to fame had not yet been established. And, you know, he himself really was a young person just starting out.

Mr. Stanley asked, "Did he say anything to you that you remember?" Mr. Myers replied:

> He made a kidding remark to me. He asked, "What are you doing this summer?" And I said I was working for the highway department. And then he said, just laughingly, "Well, we could use you on our Freedom Marches."

When asked if he thought about joining him, Mr. Myers replied, "Actually, I didn't even really know then what the Freedom Marches were. I think when you're a kid, you don't follow politics as closely as you do later."

Mr. Myers said he was cutting the grass the next day when he saw a man taking down the license plate numbers of cars in the parking lot. He asked the man what he was doing. The man said, "I'm taking pictures of the license plates on behalf of Dr. King because there's a lot of incorrigibles that latch onto his Freedom Marches." It turned out the man really was a F.B.I. agent, which continually spied on Dr. King and other civil rights leaders.

> Having met him made me more interested in his work and the things he was trying to accomplish. I wanted to read and understand more about what he was doing in the world. The fact that he was helping to make history was interesting to me because I was a history major in college. I also vividly remember that day in April 1968 when he was assassinated. I was student teaching in West Frankfort. Of course, his death seemed to have more of a personal meaning for me because I had actually met him.

The fireplace room in the Club House in 2020.

The Club House front entrance in 2020.

The beautifully restored front of the Club House interior.

The original dining hall clock and the lamp that hung over the main entrance, restored and working in the U.A.W. Club House building in 2020.

Above left: The restored banisters to the second floor of the Club House.

Above right: The former TB hut that was saved had the original door knob and lock.

Right: This is how the original hut looked when the U.A.W. found it, before restoration.

The closet and vanity in the restored tent hut in 2020.

The restored tent colony hut on the U.A.W. property today.

Starved Rock and Matthiessen state parks near Ottawa.

About the Author

JIM RIDINGS was born in Joliet, Illinois. He earned a bachelor of science degree in journalism from Southern Illinois University at Carbondale, with a minor in history. He was a reporter for *The Daily Times* in Ottawa and *The Beacon-News* in Aurora. He won more than a dozen awards for investigative reporting at both newspapers, from the Associated Press, United Press International, Copley Press, Illinois Press Association, Northern Illinois Newspaper Association, SDX Society of Professional Journalists, and other organizations.

Mr. Ridings was presented a Studs Terkel Humanities Service bronze medal from the Illinois Humanities Council in 2006. He was instrumental in getting a state historical marker erected at the site of the former coal mining town of Cardiff, Illinois.

Mr. Ridings is the author of twenty-nine books of Illinois history, including *Small Justice*; *Len Small: Governors and Gangsters*; *Cardiff: Ghost Town on the Prairie*; *Wild Kankakee*; *County West: A Sesquicentennial History of Western Kankakee County*; *The Illustrated History of the Cherry Mine Disaster of 1909*; and *The Society of the Living Dead: The Illustrated History of the Radium Dial Scandal of Ottawa*. Several of his books have won awards from the Illinois State Historical Society.

Jim Ridings with Steve Sanders of WGN-TV Channel 9 in Chicago in 2015.